Transmaterial 3

Transmaterial 3

EDITED BY
BLAINE BROWNELL

Transmaterial 3

A CATALOG
OF MATERIALS
THAT
REDEFINE
OUR
PHYSICAL
ENVIRONMENT

PRINCETON ARCHITECTURAL PRESS
NEW YORK

TO MY GRANDPARENTS, BLAINE AND ANNETTE BROWNELL

Published by
Princeton Architectural Press
37 East 7th Street
New York, New York 10003

For a free catalog of books, call 1-800-722-6657
Visit our website at www.papress.com

Images on p. 224 courtesy of Katsuhisa Kida
and Jun Takagi.

Editor: Becca Casbon
Series design: Paul Wagner
Layout: Bree Anne Apperley

Special thanks to:
Nettie Aljian, Sara Bader, Nicola Bednarek,
Janet Behning, Carina Cha, Thomas Cho,
Penny (Yuen Pik) Chu, Carolyn Deuschle,
Russell Fernandez, Pete Fitzpatrick,
Wendy Fuller, Jan Haux, Clare Jacobson,
Linda Lee, Laurie Manfra, John Myers,
Katharine Myers, Dan Simon, Andrew Stepanian,
Jennifer Thompson, Joseph Weston, and
Deb Wood of Princeton Architectural Press
—Kevin C. Lippert, publisher

Library of Congress Cataloging-in-Publication Data
Transmaterial 3 : a catalog of materials that redefine our
physical environment / edited by Blaine Brownell.
   p.   cm.
Includes indexes.
ISBN 978-1-56898-893-1 (alk. paper)
1. Materials—Technological innovations. 2. Materials—
Catalogs. I. Brownell, Blaine Erickson, 1970– II. Title:
Transmaterial three.
TA403.6.T15 2010
620.1'1—dc22
                                    2009026707

*Disclaimer: Princeton Architectural Press and the
author make no warranties or representations concerning
these products or their use.*

TABLE OF CONTENTS

INTRODUCTION   6
PRODUCT PAGE KEY   12

01: **CONCRETE**   15

02: **MINERAL**   29

03: **METAL**   47

04: **WOOD**   67

05: **PLASTIC + RUBBER**   91

06: **GLASS**   137

07: **PAINT + PAPER**   155

08: **FABRIC**   167

09: **LIGHT**   197

10: **DIGITAL**   219

DESIGNER INDEX   247
MANUFACTURER INDEX   248
PRODUCT INDEX   250
ACKNOWLEDGMENTS   252

INTRODUCTION
INNOVATION AND RESPONSIBILITY

The enthusiasm conveyed by the architecture and design communities toward emergent materials in the early 2000s was primarily a reaction to their novelty. So many new material offerings had surfaced for these and other disciplines that one could not help but take note, especially considering the degree to which they contrasted with conventional options. However, an era of novelty is giving way to one of stark realities. The acknowledgment of the severe environmental challenges humanity faces—coupled with the prominent ecological footprints of construction and manufacturing—suggests that the consideration of new, more sustainable materials is more a necessity than a luxury. However, current modes of industrial production and the energy-intensive global economy that supports them cannot be so easily retooled. The result is an unprecedented challenge—and opportunity—to redesign not only the materials used in commercial products and architecture, but also the processes by which these materials are manufactured and deployed.

The current state of the world is influenced by several important milestones that affect material development, all of which have transpired during the last few years. One milestone concerns the migration of rural populations to cities in record numbers, and our recent transformation to a predominantly urban planet. Another milestone concerns a newfound awareness about the delicate and complex interdependency of global economies; a fact made painfully evident during the recent global recession. This interdependency amplifies the difficulty of decisions related to resource allocation and use, such as the recent food vs. fuel debate that arose after agricultural products were promised for biofuel utilization. Yet another related milestone is the realization that the enterprises of humanity are no longer separate from nature, but rather influence environmental processes in profound ways. This awareness is evident in the extent to which environmental issues have become mainstream media concerns. Today, green design enjoys a ubiquity in the press unlike any previous design-related movement, and the heightened consumer awareness provides architects and designers with an unprecedented opportunity to harness innovation for the sake of environmental responsibility.

MOTIVATIONS 3.0

With so much attention being paid to greenhouse gas–related emissions and imminent fuel supply limits, energy has become a pressing topic of concern. Predictions of global peak oil (the midway point of oil depletion, beyond which oil will be more difficult to extract and thus, more expensive) inspire both the search for alternative fuel sources as well as the conservation of energy in all of its uses. Bio-based fuel sources such as ethanol, biodiesel, or biomass have begun to replace a small percentage of petroleum-derived sources. Meanwhile, renewable energy industries such as solar and wind power are propelled by increased market support and government subsidies. Predictions of future energy delivery models indicate a profound transformation from central, long distance–based service models to highly distributed, community-based frameworks. Such models will require buildings to be more active participants in energy production, and new technologies promise to add energy-harnessing functionality to architectural surfaces—thus diversifying and stabilizing a building's "energy portfolio."

Energy considerations are obviously not limited to fuel and building operation concerns; the power required during material manufacture and processing is also under scrutiny. Low-embodied-energy and zero-energy processes are slowly beginning to replace conventional methods requiring large amounts of energy. The most interesting of these involve biochemical and exothermic procedures whereby materials self-assemble according to predetermined chemical processes. Repurposing waste materials via low-energy preparations is another means for reducing energy while limiting the waste stream burden. This kind of surrogate approach is not limited to reused items; hydrocarbon-based materials are also being slowly replaced by carbohydrate-based versions due to concerns about limited oil supplies as well as the health hazards demonstrated by the processing of certain petroleum-based polymers like PVC. This post-petro-materialism offers exciting new opportunities but also raises perplexing dilemmas. As new polymers are manufactured from increasing numbers of bio-based resources, for example, the demand for agriculture and timber supplies accelerates. This demand is often in direct competition with the need for food, medicine, livestock feed, and other uses. One of the promises of bio-based materials, however, is their eventual biodegradability and reintegration into the biological nutrient cycle—as opposed to the relegation of petro-polymers to the so-called technical nutrient cycle. An interesting subset of this bio-based approach is the resurrection of vernacular material traditions into new forms that are suitable for contemporary applications.

Material manufacturing processes are also becoming increasingly sparing of water use by necessity. With accessible fresh groundwater supplies accounting for less than 1 percent of the water on the planet, manufacturers are reducing potable water consumption levels and developing new means for recycling wastewater. Moreover, unhealthy water cycles within urbanized areas have also come under increased scrutiny, and a new generation of pervious paving and geotextile systems offer improved approaches for storm water filtration and groundwater recharge.

As a matter of general practice, conservation is evident in the design thinking behind many new materials. One conservation method is represented by the optimization of material properties in terms of structural performance, largely made possible by shaping material so that it may support maximum loads over the longest spans. In this way, less material can be made to accomplish more, simply by aligning design thinking with knowledge of a material's mechanical properties. This strategy connects with the recently inspired trend of biomimicry, which takes cues from natural models for optimal formal, structural, or other performance-based outcomes. These outcomes are increasingly enabled via the use of digital fabrication technologies that are preprogrammed to optimize particular material qualities. One such approach seeks to accomplish multiple functions within a single product or system, emulating the hybrid material integrity found in many biological examples. Another approach involves the marriage of living and nonliving materials—such as green walls, planter tables, or vertical gardens—that blur disciplinary lines between architecture and landscape architecture. Further thinking about natural processes has resulted in products designed to undo the damage caused by polluting industrial practices. These so-called remediating materials photocatalyze airborne pollution and employ passive self-cleaning technologies to do more good than harm, rather than simply minimizing harm.

One of the difficulties in adopting environmentally friendly habits concerns the inability to measure the current use of energy and material resources. Smart interface technologies promise to improve this situation by augmenting building surfaces and products with distributed sensors that actively measure contextual environmental data. Responsive material systems are endowed with the ability to transform based upon shifting values beyond certain thresholds—such as architectural surfaces that flex and "breathe" in the presence of polluted air. These surfaces may also possess the ability to self-heal, especially in situations that anticipate high physical stresses or require ultralightweight components. The collective promise of such technologies is that our constructed environment will become smarter, at least in the sense that it will actively alert us to a variety of measurements that we do not currently monitor well. Building surfaces will also become more interactive, expanding interface design beyond the product scale to architectural and urban scales. One downside of ubiquitous computation is the additional power and embodied energy required by such systems, although each module within the system may be designed to harness its own energy or draw minimal electrical power.

TRENDS

As established in the first *Transmaterial* volume, several broad categories serve to elucidate current material transformations. These classifications highlight important themes shared between dissimilar products and make unexpected connections. For example, an aluminum floor system and polypropylene chair are made of different substances, but they could be similarly important in their use of recycled materials. The seven broad categories I have proposed are as follows:

1. ULTRAPERFORMING

Throughout human history, material innovation has been defined by the persistent testing of limits. Ultraperforming materials are stronger, lighter, more durable, and more flexible than their conventional counterparts. These materials are important because they shatter known boundaries and necessitate new thinking about the shaping of our physical environment.

As discussed before, the ongoing pursuit of thinner, more porous, and less-opaque products indicates a notable movement toward greater exposure and ephemerality. It is no surprise that ultra-performing materials are generally expensive and difficult to obtain, although many of these products are being developed for a broad market.

2. MULTIDIMENSIONAL

Materials are physically defined by three dimensions, but many products have long been conceived as flat surfaces. A new trend exploits the z-axis in the manufacture of a wide variety of materials for various uses, ranging from fabrics to wall and ceiling treatments. Greater depth allows thin materials to become more structurally stable, and materials with enhanced texture and richness are often more visually interesting. Augmented dimensionality will likely continue to be a growing movement, especially considering the technological trends toward miniaturization, systems integration, and prefabrication.

3. REPURPOSED

Repurposed materials may be defined as surrogates, or materials that are used in the place of materials conventionally used in an application. Repurposed materials provide several benefits, such as replacing precious raw materials with less-endangered, more plentiful ones; diverting products from the waste stream; implementing less-toxic manufacturing processes; and defying convention. A subset of this group comprises objects considered repurposed in terms of their functionality, such as tables that become light sources and art that becomes furniture.

As a trend, repurposing underscores the desire for adaptability and an increasing awareness of our limited resources. While the performance of repurposed materials is not always identical to that of the products they replace, sometimes new and unexpected benefits arise from their use.

4. RECOMBINANT

Recombinant materials consist of two or more different materials that act in harmony to create a product that performs greater than the sum of its parts. Such hybrids are created when inexpensive or recyclable products are used as filler, when a combination allows for the achievement of multiple functions, when a precious resource may be emulated by combining less-precious materials, or when different materials act in symbiosis to exhibit high-performance characteristics.

Recombinant materials have long proven their performance in the construction industry. Reinforced concrete, which benefits from the compressive strength and fireproof qualities of concrete and the tensile strength of steel, is a classic recombination. These materials are often composed of down-cycled components, which may be difficult if not impossible to re-extract, and the success of recombinant materials is based on their reliable integration, which is not always predictable. However, the continued value exhibited by many such hybrids is evidence of its growing popularity.

5. INTELLIGENT

Intelligent is a catchall term for materials that are designed to improve their environment and that often take inspiration from biological systems. They can act actively or passively and can be high- or low-tech. Many materials in this category indicate a focus on the manipulation of the microscopic scale.

Intelligence is not used here to describe products that have autonomous computational power but rather products that are inherently smart by design. The varied list of benefits provided by these materials includes pollution reduction, water purification, solar-radiation control, natural ventilation, and power generation. An intelligent product may simply be a flexible or modular system that adds value throughout its life cycle.

6. TRANSFORMATIONAL

Transformational materials undergo a physical metamorphosis based on environmental stimuli. This change may occur automatically based on the inherent properties of the material, or it may be user-driven. Like intelligent materials, transformational materials provide a variety of benefits, including waste reduction, enhanced ergonomics, solar control, illumination, as well as unique phenomenological effects. Transformational products offer multiple functions where one would be expected, provide benefits that few might have imagined, and help us view the world differently.

7. INTERFACIAL

The interface has been a popular design focus since the birth of the digital age. Interfacial materials, products, and systems navigate between the physical and virtual realms. As we spend greater amounts of time interacting with computer-based tools and environments, the bridges that facilitate the interaction between the two worlds are subject to further scrutiny.

So-called interfacial products may be virtual instruments that control material manufacture or physical manifestations of digital fabrications. These tools provide unprecedented and unimaginable capabilities, such as enhanced technology-infused work environments, rapid-prototyping of complex shapes, integration of digital imagery within physical objects, and making the invisible visible.

Interfacial materials employ the latest computing and communications technologies and suggest society's future trajectory. Interfacial materials are not infallible, but they expand our capabilities into uncharted territory.

FROM LIMITS TO OPPORTUNITIES

A basic assessment of the most innovative cultures suggests that creativity often flourishes due to—rather than in spite of—limitations. In this sense, resourcefulness is more important than resources. Therefore, the serious environmental and economic challenges currently confronting humanity may offer our greatest opportunity yet to innovate. Although the exuberance of unchecked capitalism that caused a global recession inspired a return to the "new normal," such a state need not be defined by a reduction of creativity; nor should the imminent energy, water, and materials crises cause us to withdraw from the call to scientific achievement. I have witnessed in the designers, manufacturers, and suppliers of materials represented here a persistent struggle to enhance, improve, and reshape the physical environment—often in the face of discouraging obstacles. If these individuals teach us one lesson, it is that we are at our best when we innovate.

Transmaterial is a constantly evolving project, with a growing community of material manufacturers, designers, artists, scientists, architects, builders, students, and engineers that recommend, assess, and test new products and materials. Please visit www.transmaterial.net as well as the website of Princeton Architectural Press's companion program, *Materials Monthly*, at www.materialsmonthly.com, where you may continue to discover inspiring developments in the world of materials beyond those contained within this collection.

Blaine Brownell

PRODUCT PAGE KEY

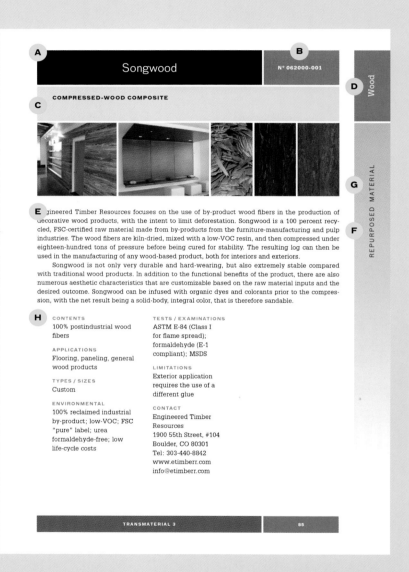

A

Songwood

B

N° 062000-001

D Wood

C

COMPRESSED-WOOD COMPOSITE

G

F

REPURPOSED MATERIAL

E Engineered Timber Resources focuses on the use of by-product wood fibers in the production of decorative wood products, with the intent to limit deforestation. Songwood is a 100 percent recycled, FSC-certified raw material made from by-products from the furniture-manufacturing and pulp industries. The wood fibers are kiln-dried, mixed with a low-VOC resin, and then compressed under eighteen-hundred tons of pressure before being cured for stability. The resulting log can then be used in the manufacturing of any wood-based product, both for interiors and exteriors.

Songwood is not only very durable and hard-wearing, but also extremely stable compared with traditional wood products. In addition to the functional benefits of the product, there are also numerous aesthetic characteristics that are customizable based on the raw material inputs and the desired outcome. Songwood can be infused with organic dyes and colorants prior to the compression, with the net result being a solid-body, integral color, that is therefore sandable.

H CONTENTS
100% postindustrial wood
fibers

APPLICATIONS
Flooring, paneling, general
wood products

TYPES / SIZES
Custom

ENVIRONMENTAL
100% reclaimed industrial
by-product; low-VOC; FSC
"pure" label; urea
formaldehyde-free; low
life-cycle costs

TESTS / EXAMINATIONS
ASTM E-84 (Class I
for flame spread);
formaldehyde (E-1
compliant); MSDS

LIMITATIONS
Exterior application
requires the use of a
different glue

CONTACT
Engineered Timber
Resources
1900 55th Street, #104
Boulder, CO 80301
Tel: 303-440-8842
www.etimberr.com
info@etimberr.com

A. NAME

The trademarked name of the particular entry being featured

B. NUMBER (Nº)

This nine-digit identification number is unique to each entry. The first six digits are based on the new MasterFormat material classification system, published June 8, 2004, by the Construction Specifications Institute. The last three digits are used to identify each product within a serial list. This numbering system is congruent with the *Materials Monthly* program, also published by Princeton Architectural Press.

C. DESCRIPTION

A brief, generic explanation of each entry

D. CATEGORY

Refers to the basic materiality of the product, such as concrete, metal, or plastic; it is the primary means of organization in this book.

E. SUMMARY

A basic text description of each entry

F. TREND

This field assigns one of the seven trends mentioned in the introduction to each entry: ultraperforming, multidimensional, repurposed, recombinant, intelligent, transformational, or interfacial.

G. TYPE

Defines each entry as a material, product, or process

H. ADDITIONAL DATA

The following information is also used to describe product entries: contents, applications, types or sizes, environmental benefits, industry tests or examinations, limitations, and manufacturer contact information.

01: **CONCRETE**

12 Blocks

MULTIFUNCTIONAL CONCRETE BLOCKS

Designed by Ralph Nelson, 12 Blocks bring innovation to a time-tested building component—the common concrete block. Each block has a unique sculptural profile that engages light, shade, and shadow and is designed to support and enhance a specific environmental condition of precipitation, sediment, growth, or habitat.

Available in a range of sizes, admixtures, and insulation inserts, the blocks are ideally suited for building, landscape, and civil-engineering construction. The 12 Blocks can be produced anywhere in the world, using primitive wet-cast forms or advanced dry-form production equipment.

CONTENTS
Concrete, with fly-ash, furnace-slag, or rice-hull pozzolan

APPLICATIONS
Building, landscape, and civil-engineering projects

TYPES / SIZES
8 x 8 x 16" (20.3 x 20.3 x 40.6 cm) standard, custom sizes available; types include Egg, Fold, Pachinko, and Ripple

ENVIRONMENTAL
Low embodied energy; use of fly-ash, furnace-slag, or rice-hull pozzolan to offset portland cement use; produced regionally, within a fifty-mile radius of a job site

TESTS / EXAMINATIONS
Prototype phase

CONTACT
Loom
708 Vandalia Street
St. Paul, MN 55114
Tel: 612-501-3983
www.loomstudio.com
hello@loomstudio.com

BubbleDeck

LIGHTWEIGHT, HOLLOW, BIAXIAL CONCRETE-SLAB SYSTEM

BubbleDeck is a structural element that virtually eliminates all concrete from the middle of a floor slab not performing any structural function, thereby dramatically reducing structural dead weight. BubbleDeck is based on a new patented technique that involves directly linking air and steel. Void formers in the middle of a flat slab eliminate 35 percent of its self-weight, removing the constraints of high dead loads and short spans. BubbleDeck's flexible layout easily adapts to irregular and curved-plan configurations. The system allows for the realization of longer spans, more-rapid and less-expensive erection, and the elimination of down-stand beams. According to the manufacturer, BubbleDeck can reduce total project costs by 3 percent.

CONTENTS
Concrete, reinforcing steel, void-forming spheres

APPLICATIONS
Structural floor slabs, shear walls

TYPES / SIZES
Wide variety of standard deck heights; can be delivered either as semiprecast filigree elements or as reinforcement modules (bubble lattice)

ENVIRONMENTAL
100% recyclable; optimal use of materials, resulting in lower embodied energy, transportation demands, and greenhouse gas emissions

TESTS / EXAMINATIONS
Fully tested in universities and test facilities in the Netherlands, Germany, and Denmark; accepted according to European codes

CONTACT
BubbleDeck
Roesevangen 8
Farum, 3520
Denmark
Tel: 0045 44-95-59-59
www.bubbledeck.com
info@bubbledeck.com

Creacrete

ULTRADENSE CONCRETE

Concrete is an omnipresent material in architecture and public spaces. Previous attempts to use it in product design have led to massive, heavy objects that were also limited in their formal design. Creacrete is a concrete-based material that is highly dense and compact and makes thin-walled objects with glossy surfaces possible for the first time. Creacrete explores new sides of concrete with the aim to realize surfaces that are permanently glossy, abrasion- and acid-resistant, food-safe, and hydrophobic.

Developed by Alexa Lixfeld, Creacrete shows that concrete can be an alternative material for ceramics. The use of concrete shortens the production process compared to that of ceramics because it eliminates the need for two kiln firings—thus reducing energy input and expenses. The simplified production process also allows new freedom in form-making. Just as ceramics need a glaze coating to be resistant to stains, concrete needs adequate treatment to make it stain-resistant and food-safe.

CONTENTS
Concrete

APPLICATIONS
Floor and wall coverings, tiles, sanitary objects, decorative objects, facades, tableware

TYPES / SIZES
Custom colors and finishes; unit thickness from .12" (3 mm)

ENVIRONMENTAL
Reduced CO_2 output

LIMITATIONS
Not as heat-resistant as ceramics

CONTACT
AlexaLixfeld Design GmbH
Bahrenfelder Steindamm 39
Hamburg, 22761
Germany
Tel: +49 (0)172-5114671
www.creacrete.com
info@alexalixfeld.com

INTEGRAL CONCRETE WATERPROOFING ADMIXTURE AND CORROSION INHIBITOR

Hycrete's Element is an environmentally friendly admixture that integrally waterproofs concrete used in commercial construction. Certified Cradle to Cradle by McDonough Braungart Design Chemistry (MBDC), Element eliminates the need for external membranes typically used to waterproof concrete, thereby making the concrete more easily recyclable following demolition. This approach can eliminate thousands of pounds of volatile organic compounds (VOCs), CO_2, and non-renewable content. Additionally, the admixture enhances structure durability by protecting against corrosion of steel rebar.

In eliminating the need for a manually applied membrane, Hycrete Element can also save time in construction schedules. With typical membrane applications, contractors must often wait for the concrete to dry before a waterproofing subcontractor can apply the membrane—even after rainfall and rewetting. In contrast, Hycrete Element is dosed during concrete mixing and is not subject to weather delays.

Hycrete Element is available with the IntegraTek warranty, which is an industry best-in-class, ten-year warranty for waterproofing.

CONTENTS
Cradle to Cradle–certified polymer, registered as a biological nutrient

APPLICATIONS
Retaining walls, plaza decks, pools and ornamental fountains, green roofs, bridge decks, tunnels, water treatment facilities, piers and docks, parking garages (elevated and below grade), airport runways and aprons, certain industrial floor applications

TYPES / SIZES
Dosage for waterproofing is one U.S. gallon per cubic yard of concrete (five liters per cubic meter)

ENVIRONMENTAL
Cradle to Cradle certified, allows Hycrete-treated concrete to be more easily recycled by replacing external membranes

TESTS / EXAMINATIONS
British Standard Institute 1881.122

LIMITATIONS
For concrete applications where waterproofing and/or corrosion inhibition are design parameters

CONTACT
Hycrete, Inc.
462 Barell Avenue
Carlstadt, NJ 07072
Tel: 201-763-9227
www.hycrete.com
aayer@hycrete.com

fibreC

GLASS FIBER REINFORCED CONCRETE (GFRC)

FibreC is a lightweight, high-performance concrete that requires no steel reinforcement. A special extrusion process incorporates layers of fiberglass into a concrete matrix; in the top layer and underlayer, the fibers are undirected and scattered, and in the medium layer they take the form of roofings (fiber bundles). The omission of steel reinforcement allows the construction of slim concrete elements that are highly stressable despite being extremely thin-walled. The result is a slab .3 to .5 inches (8 to 13 millimeters) thick, which is very lightweight, yet has a high flexural strength. FibreC slabs are fabricated in different colors before being hardened for twenty-eight days. Owing to its formability, the so-called concrete skin offers flowing transitions from interior to exterior surfaces and a smooth covering for edges and corners. As fibreC can be used for all surfaces, traditional spatial boundaries disappear, and interior and exterior conditions may be treated similarly.

CONTENTS
90% sand and cement, 10% glass fiber, pigments, and different agents

APPLICATIONS
Concrete applications that require increased fire resistance, highest loading capacity at minimum cross sections, long-term durability, and formability

TYPES / SIZES
Exterior sizes: 47 5/8" x 8'-2 27/64" x 3/8" (121 x 250 x 1 cm), 47 5/8" x 8'-2 27/64" x 1/2" (121 x 250 x 1.3 cm), 47 5/8" x 11'-9 3/4" x 3/8" (121 x 360 x 1 cm), 47 5/8" x 11'-9 3/4" x 1/2" (121 x 360 x 1.3 cm); interior sizes:

47 5/8" x 8'-2 27/64" x 5/16" (121 x 250 x .8 cm), 47 5/8" x 8'-2 27/64" x 1/2" (121 x 250 x 1.3 cm), 47 5/8" x 11'-9 3/4" x 3/8" (121 x 360 x 1 cm), 47 5/8" x 11'-9 3/4" x 1/2" (121 x 360 x 1.3 cm)

ENVIRONMENTAL
Efficient use of material, food-safe (used in ovens)

TESTS / EXAMINATIONS
35 international product and system tests, including: DIN EN ISO 9001, DIN EN ISO 14001, ETA, IBO, ÖNORM EN 12467, ASTM

LIMITATIONS
Exterior: restrictions on application for 3/8" (10 mm) panel; interior: restrictions on application for 5/16" (8 mm) and 3/8" (10 mm) panel

CONTACT
Rieder Faserbeton-Elemente GmbH
Glasberg 1
Kolbermoor, 83059
Germany
Tel: +49 (0)8031-90167-0
www.rieder.cc
office@rieder.cc

LIGHT-SENSITIVE INTERFACE INTEGRATED WITHIN A CONCRETE SURFACE

Light-Sensitive Concrete is a technology that allows concrete to be sensitive to ambient light levels. It senses the luminosity distribution throughout a concrete surface and sends the data to a computer. By converting this data to various values within custom-designed software, one can control sound, light, projected visuals, and other effects by modifying the light condition on the concrete surface.

Developed by University of Tokyo researcher Tokihiko Fukao, Light-Sensitive Concrete consists of concrete, embedded optical fibers, photodiodes, and electrical circuitry. Optical fibers are distributed within a regular grid, and sensors are attached beneath them in the same arrangement. The interactive properties of the material are intentionally hidden within what appears to be conventional concrete—suggesting possibilities for other light-sensitive building materials and surfaces as part of a total ambient interactive system.

CONTENTS
Cement, aggregates, optical
fibers, photodiodes,
electronics

APPLICATIONS
Furniture, architectural
installations, interactive art

TYPES / SIZES
Tabletop 17.7 x 53.2 x 2.8"
(45 x 135 x 7 cm); custom
sizes available

LIMITATIONS
Sensitive to ambient
brightness

CONTACT
Noguchi Laboratory,
University of Tokyo
T-building No.1,
Hongo 7-3-1
Tokyo, 113-8656
Japan
Tel: +81 3-5841-6202
http://bme.t.u-tokyo.ac.jp
webmaster@bme
.arch.t.u-tokyo.ac.jp

INTERFACIAL MATERIAL

PIXA

CONCRETE VIDEO SCREEN

PIXA, designed by Abhinand Lath, utilizes Sensitile Systems' light-piping technologies to enable the placement of a precise grid of light guides within a concrete matrix. These light guides optically connect the two surfaces of the material and mate directly to an LED screen or projection system on one surface, allowing the digital control of each "pixel" on the corresponding surface. PIXA is also appropriate for use as a screen or divider panel, and may be manufactured with a random, "organic" pattern of light terminals rather than a strict grid layout.

CONTENTS
Polymer-modified cementitious panel

APPLICATIONS
Floor or wall/facade video screens, privacy screens, light diffusers

TYPES / SIZES
23.6 x 23.6 x 1.5" (60 x 60 x 3.8 cm) overall with .12 x .12" (3 x 3 mm) light terminals placed 1" (2.5 cm) on center (other sizes and light terminal patterns available)

ENVIRONMENTAL
Low energy use

TESTS / EXAMINATIONS
ASTM D695, D638, E84; others available upon request

LIMITATIONS
3/4" (1.9 cm) thickness limitation

CONTACT
Sensitile Systems
1735 Holmes Road
Ypsilanti, MI 48198
Tel: 313-872-6314
www.sensitile.com
info@sensitile.com

ANTICORROSION, ANTICRACKING CONCRETE

The making of cement—the main ingredient in concrete—produces 8–10 percent of the world's CO_2. In fact, making one ton of cement generates one ton of CO_2 gas. Architect and researcher Carolyn Dry has developed a self-repairing cement in order to expand the longevity of concrete. Conventional concrete's porosity makes it vulnerable to environmental intrusions. However, Dry's new cement makes Self-Repairing Concrete less porous, and also prevents corrosion and cracking.

CONTENTS
Cement

APPLICATIONS
Concrete bridges, beams, deck plates, pavement

TYPES / SIZES
Custom

ENVIRONMENTAL
Enhances durability of concrete

TESTS / EXAMINATIONS
Compression and flexure tests, corrosion testing, in-field tests for flexure, fiber optics and ETDR tests

LIMITATIONS
Added cost

CONTACT
Natural Process Design Inc.
1250 East 8th Street
Winona, MN 55987
www.naturalprocessdesign
.com
naturalprocessdesign@
yahoo.com

ULTRAPERFORMING MATERIAL

Terrazzo Lumina

ILLUMINATED CONCRETE SURFACES

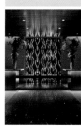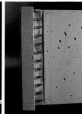

Sensitile Terrazzo Lumina slabs and tiles, designed by Abhinand Lath, are refined concrete surfaces designed to propagate and emit light. Illumination travels within the concrete via embedded light-guides and is emitted in pixelated form. The manufacturing process is quite flexible and various geometries and configurations of light terminals are possible. The placement of the light source relative to the emitting surface is also flexible (and serviceable), making the product versatile for use in bar tops, flooring, wall features, and infrastructural projects.

CONTENTS
Polymer-modified concrete, LEDs

APPLICATIONS
Vertical and horizontal surfacing

TYPES / SIZES
Slabs made to custom sizes up to 8' (2.4 m) long; standard tiles 24 x 24 x 1 1/2" (61 x 61 x 3.8 cm) or 24 x 48 x 1 1/4" (61 x 122 x 3.2 cm)

ENVIRONMENTAL
Allows the strategic illumination of concrete using high-efficiency LED technologies

TESTS / EXAMINATIONS
ASTM D638, D695, E84; others available upon request

LIMITATIONS
Presently cannot be made thinner than 1 1/4" (3.2 cm)

CONTACT
Sensitile Systems
1735 Holmes Road
Ypsilanti, MI 48198
Tel: 313-872-6314
www.sensitile.com
info@sensitile.com

CONCRETE BUILDING-INSULATION SYSTEM

The Thermomass building system is a pre-engineered, energy-efficient insulated concrete sandwich-wall system developed by Composite Technologies Corporation. Thermomass is made with high-strength fiber-composite connectors and rigid insulation, which is designed to be sandwiched between two layers of concrete. The three resulting layers are held together by connectors on a 12-inch (30.5-centimeter) or 16-inch (40.6-centimeter) grid. The system may be applied to tilt-up, precast, and cast-in-place concrete panels.

CONTENTS
Fiber-composite connectors, rigid insulation

APPLICATIONS
Site-cast tilt-up, plant-precast, and poured-in-place concrete sandwich wall construction

TYPES / SIZES
Noncomposite panels, composite panels, extruded polystyrene insulation, polyisocyanurate insulation; R-values from R-5 to R-50

ENVIRONMENTAL
High thermal efficiency, high content of recycled and regional materials

TESTS / EXAMINATIONS
ICC-ES (ESR 1746) and City of Los Angeles Research Report

LIMITATIONS
Requires installation by precertified concrete contractors

CONTACT
Composite Technologies Corporation
1000 Technology Drive, P.O. Box 950
Boone, IA 50036
Tel: 800-232-1748
www.thermomass.com
info@thermomass.com

Xposed

CEMENT-BASED ECO-COMPOSITE

  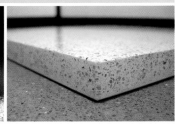

Xposed is a refined, cement-based eco-composite made from natural minerals and up to 25 percent postindustrial recycled content. Standard slab materials are available and can be fabricated into countertops, tabletops, surrounds, or display tops using ordinary natural stone fabrication tools and equipment. Xposed is also available for custom interior and exterior applications such as furniture elements, sinks, tiles, and three-dimensional objects. Twelve standard colors are available as well as custom colors.

CONTENTS
Silica sand, cement, postindustrial recycled material, fibers

APPLICATIONS
Countertops, tabletops, tiles, furniture, sinks

TYPES / SIZES
Standard prefabricated slabs 30 x 96 x 1.5" (76.2 x 243.8 x 3.8 cm); custom shapes and sizes possible

ENVIRONMENTAL
Up to 25% postindustrial recycled content, recyclable

LIMITATIONS
Nonstructural

CONTACT
Meld
3001-103 Spring Forest Road
Raleigh, NC 27616
Tel: 919-790-1749
www.meldusa.com
info@meldusa.com

02: **MINERAL**

Cassini Blocks

KIDNEY-SHAPED CERAMIC BUILDING BLOCKS

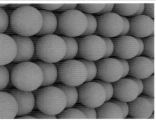  

Designed by Neil Meredith as a new type of wall system, Cassini Blocks embrace the natural and formal tendencies of ceramics as a starting point for design. Utilizing a single unit type shaped like a kidney or peanut, many different stacking and opening possibilities emerge. Instead of forcing the material into a standard brick unit, these building units are slip-cast into hollow blocks. Ceramic building units also offer distinct performance criteria: clay is a durable and natural material, has a high insulation capacity, absorbs and distributes water when uncoated, and can shed water once glazed—allowing it to be used in interior and exterior applications.

CONTENTS
Slip-cast red earthenware clay

APPLICATIONS
Interior and exterior walls

TYPES / SIZES
15 3/4 x 7 1/4" (40 x 18.4 cm) individual blocks; stacked dimensions vary

ENVIRONMENTAL
Natural material, biodegradable and reusable; requires no finishing or maintenance

LIMITATIONS
Large installations may require additional engineering and testing

CONTACT
Sheet Design
2134 45th Road, #16
Long Island City, NY 11101
Tel: 313-429-0615
www.sheetd.com/loosefit
info@sheetd.com

CARBON FIBER–REINFORCED CERAMIC COMPOSITE

 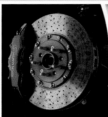  

Brembo manufactures advanced ceramic-composite braking systems for automotive applications and also performs general research and development for innovative materials. Originally developed for aerospace applications due to its light weight yet resilience under heavy loading conditions, a combination of chopped carbon fiber and silicon carbide (SiC) matrix endows the Ceramic Composite Material (CCM) with exceptional physical characteristics. Since the carbon fibers are embedded in a SiC matrix, they provide elasticity and damage tolerance while the matrix provides wear- and oxidation-resistance. The properties of extreme damage tolerance, low elastic modulus, and superior wear resistance make CCM an ideal brake disc material. CCM offers benefits with regard to weight, comfort, corrosion resistance, and long life, and may be used in dry and wet conditions.

CONTENTS
Silicon metal (Si), silicon carbide (SiC), carbon fibers

APPLICATIONS
Braking system for high-performance automotive applications

TYPES / SIZES
ø 13–16.2" (33–41 cm); thickness 1.3–1.7" (3.2–4.2 cm)

ENVIRONMENTAL
No pollution due to negligible wear during lifetime, ceramic material is completely recyclable

TESTS / EXAMINATIONS
Bench simulation

LIMITATIONS
Maximum operative temperature 2372°F (1300°C)

CONTACT
Brembo
Viale Europa 2
Stezzano, Bergamo I24040
Italy
Tel: +39 035-605-111
www.brembo.com
Direzione_BU_Auto@brembo.it

RECOMBINANT MATERIAL

Eco-Essence

NATURAL SHELL PANELS

Eco-Essence is a decorative surface panel, handmade from natural mother-of-pearl or Capiz shells, available on a selection of backings for ease of installation. Its inherent opalescence and iridescent coloring create a modern, luxurious quality that is suitable for focal walls and feature elements in many environments. This organic material is derived from the managed harvest of shells, and natural variations in texture and color may occur. The collection is available in eighteen standard patterns with custom capabilities.

CONTENTS
Capiz shells, mother-of-pearl; backings include mesh, medium-density fiberboard (MDF), ceramic, and nonwoven cloth

APPLICATIONS
Decorative surfacing applications in all commercial environments, including walls, ceilings, and column enclosures

TYPES / SIZES
Tiles 12 x 12" (30.5 x 30.5 cm) nominal

ENVIRONMENTAL
Made from the managed harvest of shells

LIMITATIONS
Interior use only

CONTACT
Architectural Systems, Inc.
150 West 25th Street,
8th Floor
New York, NY 10001
Tel: 800-793-0224
www.archsystems.com
sales@archsystems.com

Electroboard

ELECTRO-CONDUCTIVE GYPSUM WALLBOARD

Electroboard is an interior finish material that pairs flatwire technology with a fire-resistant gypsum core. It provides an electrified, low-voltage surface accessible with a proprietary connector that one pushes into the face of the wall. As an alternative to traditional track-based lighting systems, Electroboard offers unlimited fixture-configuration options and may be finished to blend seamlessly with conventional gypsum wallboard surfaces.

Designed by Eric Olsen with Superficial Studio, Electroboard is economical like conventional gypsum-based cladding and creates additional value though embedded conductivity. It offers material and energy efficiencies as well as infrastructural flexibility. The low-voltage power supplied by Electroboard is compatible with many technologies, like liquid crystal displays and OLED light fixtures, and point-of-use electrical transformers are eliminated with Electroboard—thus creating considerable energy savings.

CONTENTS
Gypsum, flatwire film, paper, electrical transformer

APPLICATIONS
Dry location wall and ceiling surfaces where low-voltage power is needed

TYPES / SIZES
4' x 8' x 1/2" (1.2 m x 2.4 m x 1.3 cm) sheet; custom sizes possible

ENVIRONMENTAL
Made from recovered manufacturing waste and recycled paper, low-voltage power supply

TESTS / EXAMINATIONS
UL listing pending

LIMITATIONS
Not for use in exterior or wet location applications

CONTACT
Superficial Studio
3163 North Beachwood Drive
Hollywood, CA 90068
www.superficialstudio.com
info@superficialstudio.com

Flexipane

FLEXIBLE, NATURAL SHELL-SURFACE COVERING

 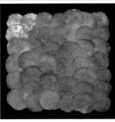  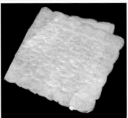

Flexipane is a seamless surface covering made of natural Capiz shells. The shells become soft and flexible after undergoing a special mechanical process. This surface covering can be used for architectural woodwork, furniture, interior paneling, countertops, and other interior applications in which wood products are traditionally used. Moreover, unlike typical surface coverings made of shells that are usually hard, Flexipane is characteristically soft and flexible—allowing easier installation on curved or rounded surfaces.

CONTENTS
Natural shells, polyester fiber, nitocellulose laquer

APPLICATIONS
Interior wall coverings, furniture

TYPES / SIZES
11.8 x 11.8 x .05" (30 x 30 x .12 cm), 11.8 x 23.6 x .05" (30 x 60 x .12 cm) panels

ENVIRONMENTAL
Widely available resource (not endangered)

TESTS / EXAMINATIONS
ASTM E84

LIMITATIONS
Not applicable for areas with direct sunlight and with high moisture such as the kitchen and bathroom

CONTACT
La Casa Deco
San Diego Street
Valenzuela City,
Metro Manila 1443
Philippines
Tel: +632 2771043 to 45
www.lacasadeco.com
info@lacasadeco.com

FLY ASH BUILDING COMPONENTS

 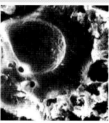 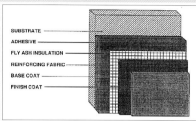

SUBSTRATE
ADHESIVE
FLY ASH INSULATION
REINFORCING FABRIC
BASE COAT
FINISH COAT

REPURPOSED MATERIAL

Fly ash, the waste product of burning coal, is largely composed of carbon and contains many heavy metals. It is often combined with cement as an additive, but only up to a certain amount.

Carolyn Dry has developed a method of fabricating building panels and insulation out of nearly 100 percent fly ash in order to sequester these heavy metals so that they do not leach out and pollute the environment. Essentially cooking the ash into a solid, Dry utilizes a flux that allows processing at lower temperatures—thus using less energy and fewer chemicals. Components such as building panels, bricks, and insulation may be produced without the need for binders such as cement.

CONTENTS
Postindustrial fly ash

APPLICATIONS
Sandwich panels, CMU substitution, exterior insulation and finish systems (EIFS) alternative

TYPES / SIZES
Custom

ENVIRONMENTAL
Sequesters heavy metals, reuses postindustrial waste

TESTS / EXAMINATIONS
Compression, flexure, water absorption, insulative properties, and chemical leaching testing and microscope analysis

LIMITATIONS
Caution required, as the material contains heavy metals

CONTACT
Natural Process Design Inc.
1250 East 8th Street
Winona, MN 55987
www.naturalprocessdesign.com
naturalprocessdesign@yahoo.com

Found Space Tiles

PROFILED CERAMIC WALL TILES

Found Space Tiles transform the common into the uncommon. Designers Stephanie Davidson and Georg Rafailidis developed the tiles after seeking greater possibilities within a conventional tiled surface. Much like appropriating found objects, these tiles are made using the found spaces between bodies and architectural surfaces, turned into positive forms. The resulting tiles are a formal hybrid between two very necessary and basic architectural elements—the body and the wall. Part body and part wall, the tiles echo the presence of a person, a posture, and literally reach out to be touched.

Made in posture- and/or body-specific clusters, the tiles are designed in standard finished dimensions, intended to be incorporated into inexpensive, tiled surfaces.

CONTENTS
Earthenware ceramic

APPLICATIONS
Wall surface finish

TYPES / SIZES
6 x 6" (15.2 x 15.2 cm) with variable depth, to be integrated with standard tiles, designed in clusters of two or more

ENVIRONMENTAL
Low-fired ceramic, low maintenance, durable

CONTACT
Touchy-Feely
Ossastrasse 15
Berlin, D-12045
Germany
Tel: +49 (0)3026343990
www.touchy-feely.net
info@touchy-feely.net

CONCRETE FORM MASONRY UNITS

The OneStep Building System is a holistic, high-performance building product that addresses the entire exterior structure of a wall. The system's concrete form masonry units (CFMUs) allow the shell of a building to be constructed using only one product, one trade organization, and one step in the construction sequence. CFMU combines the texture of masonry, the strength of cast-in-place masonry, and the energy efficiency of polystyrene insulation into a singular building product—dramatically simplifying the construction process.

The OneStep Building System incorporates high R-values, low conductivity, and dense thermal mass—all of which combine to significantly reduce heating and cooling costs. The CFMU includes a five-layer moisture-blocking system, thus preventing water leakage, condensation, and black mold. The product is available with standard masonry finishes for both the interior and exterior faces of the wall, allowing for a wide variety of combinations of colors and finishes. In addition, the CFMU installs quickly and eliminates many of the products and trades commonly required in high-performance construction.

CONTENTS
Cement, aggregates, plastic, rigid insulation

APPLICATIONS
Exterior building shells

TYPES / SIZES
Standard masonry sizes, colors, and finishes

ENVIRONMENTAL
Reduced waste, reduced energy consumption, high recycled content, reduced embodied energy

TESTS / EXAMINATIONS
ASTM E72, C740, C426, C780, C1019, Ct3L4, E514

LIMITATIONS
Not suitable for lightweight construction

CONTACT
Pentstar
1100 North Shore Drive West
Orono, MN 55364
Tel: 763-566-5400
www.pentstar.com
john@pentstar.com

RECOMBINANT PRODUCT

PS Photo Tiles

HIGH-RESOLUTION PHOTOGRAPHIC TILES

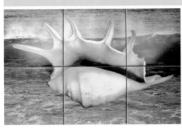

Sweden-based Parisima Kadh AB has developed a method for applying photographic images and other visual content to ceramic tiles. Unlike similar processes that offer images at the size of a tile, PS Photo Tiles allow the creation of one large, seamless image over the span of an entire tiled wall or floor surface. Parisima will produce tiles from customer-delivered content, or designers may select from more than five hundred stock pictures.

CONTENTS
Ceramic tiles

APPLICATIONS
Interior or exterior
horizontal and vertical
applications

TYPES / SIZES
From mosaic tiles up to 9.8
x 15.7" (25 x 40 cm)

TESTS / EXAMINATIONS
Water and chemical (graffiti
cleaning) tested

LIMITATIONS
Should not be scraped with
sharp objects

CONTACT
Parisima Kadh AB
Garavregatan 9A
Karlstad, Värmland 652 20
Sweden
Tel: +46 54-10-07-60
www.kadh.se
pasris20@hotmail.com

MULTIFUNCTIONAL CERAMIC BUILDING BLOCKS

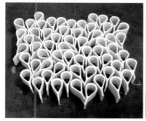   

Bricks and tiles, though materially similar, are commonly understood as discrete building materials. Bricks are stacked to produce a solid volume, while tiles are arrayed to create a surface. Q Blocks, designed by Neil Meredith, work as a single ceramic building unit that can perform a host of different constructive functions. The units can stack to create walls, overlap to create roof tiles, or nest to create irregular tiling patterns. By extruding clay through a steel die and then changing the length of the resulting tube, a single hollow profile can be used to create a range of constructive units. This approach allows for a number of advantages over the standard rectangular stacked units. For example, the air cavity within the units also allows services to pass through what would typically be a solid cross section. These bricks, pavers, and roof tiles can nest and stack in both regular and irregular patterns depending on use and installation.

CONTENTS
Extruded clay

APPLICATIONS
Interior and exterior walls and roofs

TYPES / SIZES
2 x 4 x 8/16/32" (5.1 x 10.2 x 20.3/40.6/81.2 cm) individual blocks

ENVIRONMENTAL
Biodegradable, reusable, no finishing required

LIMITATIONS
Large installations may require additional engineering and testing

CONTACT
Sheet Design
2134 45th Road, #16
Long Island City, NY 11101
Tel: 313-429-0615
www.sheetd.com/loosefit
info@sheetd.com

Sensitive Apertures

CELLULAR CERAMIC LIGHT APERTURES

Sensitive Apertures is a modular slip-cast ceramic building skin designed to admit a small quantity of light through a refractive glass aperture. This opening redirects sunlight onto the inside surface of the cell, projecting an even, luminous glow to the interior space. The 1 percent open apertures admit levels of light desired for interior circulation zones as well as insulate a building from solar heat gain and loss.

CONTENTS
Ceramic, glass

APPLICATIONS
Building skins

TYPES / SIZES
5" (12.7 cm) tetrahedron

ENVIRONMENTAL
Reduces heat gain/loss

LIMITATIONS
Aperture modules are not load bearing

CONTACT
Ben McDonald
3833 Goldwyn Terrace
Culver City, CA 90232
www.benarimcdonald.com/ceramic.html
sensitiveapertures@gmail.com

OXIDE-COATED WATER FILTRATION MEDIA AND POROUS CONCRETE MATRICES

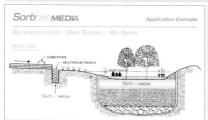

Many lakes, ponds, and rivers suffer from algae blooms and depleted dissolved oxygen as a direct result of excessive nutrients entering waterways after rainstorms. Phosphorus is the accelerant that causes algae blooms to grow rapidly. SorbtiveMEDIA is an oxide-coated water filtration system engineered to remove pollutants including phosphorus. It is produced from natural and recycled aggregates, and is applied in a wide variety of low-impact development landscape applications.

SorbtiveMEDIA is an appropriate solution for impaired watersheds or protecting water resources, with proven phosphorus-treatment performance exceeding current North American regulations. Available in a variety of gradations or as porous concrete products, SorbtiveMEDIA can accommodate many landscape designs and stormwater treatment systems. SorbtiveMEDIA can be flexibly applied to sand filters, infiltration trenches, and filter cartridges, and is also a perfect match for low impact development (LID) applications such as bioretention cells, rain gardens, gabion walls, porous-concrete pavement, and permeable-pavers systems. SorbtiveMEDIA is a safe, nonhazardous material that does not decompose or leach captured pollutants.

CONTENTS
Natural aggregates, up to 75% recycled aggregates, oxide coating

APPLICATIONS
LID and landscape-design applications used to treat and manage stormwater, such as: bioretention cells, rain gardens, gabion walls, sand filters, infiltration trenches, dry wells, permeable-pavers systems, porous concrete, and other filter cartridges

TYPES / SIZES
.006–.2" (.15–4.75 mm) filtration particle size; a variety of gradations are available

ENVIRONMENTAL
Removes phosphorus and other pollutants from stormwater, abundant natural and/or recycled raw materials, nontoxic, no VOCs

TESTS / EXAMINATIONS
Independent field performance and laboratory benchmark testing; 10 years of university research prior to commercialization

LIMITATIONS
Sorbability of dissolved phosphorus is finite, requiring calculations to determine longevity prior to exhaustion and replenishment

CONTACT
Imbrium Systems
9420 Key West Avenue, Suite 140
Rockville, MD 20850
Tel: 888-279-8826
www.imbriumsystems.com
info@imbriumsystems.com

Trigger Point

THERAPEUTIC PLASTER MOLDINGS

Massage therapists commonly work to access and manipulate multiple trigger points in the body as a means of aiding diverse ailments. Touchy-Feely's Stephanie Davidson and Georg Rafailidis consulted with massage therapists to develop forms of an appropriate depth and shape to facilitate self-trigger point massage within architectural surfaces. Trigger Point moldings are rounded fibrous plaster forms that can be integrated into a wall surface. As suggestive protrusions, the moldings encourage heightened physical interactions between bodies and architectural surfaces, and suggest that buildings can participate in the necessary work of massage therapists.

The seemingly sculpted forms of Trigger Point moldings were derived from a process of casting plaster using fabric formwork. The bulges are not designed, but rather directly capture the material behavior of the plaster poured into sewn fabric forms. The moldings can be installed easily into any new or existing drywall or plastered surface at any desired height or in any density/pattern. The moldings can also be installed with heating elements cast-in from the back, warming the forms to body temperature and assisting further in muscle tension relief.

CONTENTS
Fiber-reinforced plaster

APPLICATIONS
Tactile surface, therapeutic wall, decorative element

TYPES / SIZES
Variable depths from 2.4–4.7" (6–12 cm), with a base of 5.9 x 5.9" (15 x 15 cm)

ENVIRONMENTAL
Improves occupant well-being

CONTACT
Touchy-Feely
Ossastrasse 15
Berlin, D-12045
Germany
Tel: +49 (0)3026343990
www.touchy-feely.net
info@touchy-feely.net

PROFILED STONE TILE MOSAIC

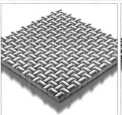   

Woven Stone is part of the ASI Specialty Products Magna Mosaic collection. Earth tones combined with dimensional, geometric features create a powerful and sculptured look, lending an artistic element to any space. The continuous, controlled texture of Woven Stone creates a sense of movement, suitable for the most modern of interiors, including retail, hospitality, healthcare, and corporate projects.

CONTENTS
Stone

APPLICATIONS
Wall surfaces, bar fronts, flooring, borders and trim, spa and wet areas

TYPES / SIZES
Tiles standard 12 x 12" (30.5 x 30.5 cm); approximate thickness 1/4–1" (6–25 mm)

LIMITATIONS
Not for structural use

CONTACT
Architectural Systems, Inc.
150 West 25th Street,
8th Floor
New York, NY 10001
Tel: 800-793-0224
www.archsystems.com
sales@archsystems.com

MULTIDIMENSIONAL MATERIAL

Y Blocks

CELLULAR CERAMIC BUILDING BLOCKS

Y Blocks, designed by Neil Meredith, are a new type of brick system that creates an informally arranged wall or paving surface while utilizing a single repeated unit. A short version can be used horizontally as paving or vertically to create a screen wall. A longer version stacks vertically to create a brick wall with the ends perpendicular to the wall surface, similar to masonry header coursing. Unlike standard bricks or tiles, there is no idealized configuration. The introduction of each new unit ripples throughout the system, letting forces internal to the stack or field determine the final design. While the system has some strange requirements—namely buttressing and patience on the part of the installer—it does solve some problems inherent with typical masonry construction, such as dealing with irregular edges and ways to introduce irregular patterns and openings.

CONTENTS
Extruded clay

APPLICATIONS
Interior and exterior walls

TYPES / SIZES
4 x 3.7 x 2/4/8/16" (10.2 x 9.3 x 5.1/10.2/20.3/40.6 cm) individual blocks; stacked dimensions vary

ENVIRONMENTAL
Biodegradable, reusable, no finishing required

LIMITATIONS
Large installations may require additional engineering and testing

CONTACT
Sheet Design
2134 45th Road, #16
Long Island City, NY 11101
Tel: 313-429-0615
www.sheetd.com/loosefit
info@sheetd.com

03: **METAL**

Adaptive Shading

GRID-BASED ADAPTIVE SHADING SYSTEM

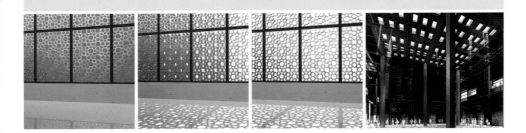

For an architectural project located in a desert climate, Hoberman Associates was asked to design a roof covering for an open-air plaza. They created a kinetic design that works off an operable grid. In its covered configuration, the shading roof appears similar to a coffered ceiling; yet when retracted, the roof becomes a slender lattice. In addition to shading, the system is designed to protect visitors from high winds and dust storms.

Each unit is a self-contained, framed grid assembly consisting of multiple panels that extend and retract synchronously. Each whole unit actually comprises several openings that are operated by a single drive arm, making the system efficient and economic.

The grid shade's simple, robust design provides opportunities for custom profiles and patterning. Hoberman has developed parametric design tools for designing customized configurations.

CONTENTS
Machined aluminum frame and slats, sand-evacuating bearings

APPLICATIONS
Adaptive building facade and roof coverings for commercial, industrial, or residential sectors

TYPES / SIZES
Custom profiles, sizes, and compositions

ENVIRONMENTAL
Controls solar gain and light levels, resulting in decreased energy consumption and improved occupant comfort; environmental barrier protects against wind and inclement weather

CONTACT
Hoberman Associates
40 Worth Street, Suite 1680
New York, NY 10013
Tel: 212-349-7919
www.hoberman.com
associates@hoberman.com

EXPANDED METAL MESH

Amico's APEX expanded metal mesh offers many benefits such as texture, passage of light, air movement, reduction of solar gain, a high strength-to-weight ratio, and a variety of manufacturing material options. Carbon steel, galvanized steel, aluminum, and stainless steel are commonly used to make expanded mesh, and Amico can also expand alloys such as brass, copper, COR-TEN, and titanium. One of the most striking aspects of expanded mesh is the small amount of raw material required to produce a large amount of product. The expanding method is a slitting and stretching process, which creates a product that is stronger and lighter than its original form and that will not unravel. Amico engineers can custom engineer new mesh designs based on functional requirements.

CONTENTS
Carbon steel, stainless steel, aluminum, copper, brass, titanium, Cor-ten, and many more alloys

APPLICATIONS
Facades, sunscreens, balustrades, wall treatments, walkways, fences, growing walls

TYPES / SIZES
1 x 1' (.3 x .3 m) up to 8 x 10' (2.4 x 3 m); custom sizes and shapes are easily produced

ENVIRONMENTAL
Efficient use of material: one 4 x 8' (1.2 x 2.4 m) sheet of raw material yields three 4 x 8' sheets of expanded mesh; recycled content

TESTS / EXAMINATIONS
EMMA 557-99, ASTM F1267-01

CONTACT
Amico
1080 Corporate Drive
Burlington, ON L7L 5R6
Canada
Tel: 289-313-2211
www.amico-online.com
pshevchenko@gibraltar1.com

MULTIDIMENSIONAL PRODUCT

AL SURFACES

Metal

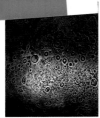

MULTIDIMENSIONAL PROCESS

Based Upon's Richard Abell and Ian Abell have developed a series of multilayered, textured surfaces made of metals, resins, and lacquers that can be specified as surface finishes within architectural and interiors projects. Taking inspiration from elements as diverse as stingray skin, laburnum leaves, or the fine lines on a human hand, Based Upon has conjured a rich and varied collection of surface features using construction-ready materials. Based Upon Surfaces may be hung like wall paintings, although they are often used to create detailed feature walls or monolithic furniture pieces. The surfaces also make visually stimulating ceilings or durable metal floors that subtly evolve over time. Working the metal from a liquid to a solid, Based Upon plays with the changing state of the material and experiments with the way the liquid metal dries, drips, or settles upon a surface. The finishing process is careful, considered, and intricate. Sanding and polishing excavates the metal from beneath the earthlike surface that forms once the metal cools, allowing the application of traditional metal techniques.

CONTENTS
Semiprecious metals, resins, and colored lacquers

APPLICATIONS
Interior fixtures and fittings, furniture, art installations, residential, commercial and hospitality, flooring

TYPES / SIZES
Surfaces adapted to any size

TESTS / EXAMINATIONS
Fire Category 0, certified for use aboard aircraft

LIMITATIONS
Hand-based techniques

CONTACT
Based Upon
Unit A121, Faircharm,
8-12 Creekside
London, SE8 3DX
United Kingdom
Tel: +44 (0)20-8320-2122
www.basedupon.co.uk
info@basedupon.co.uk

CarbonCoat

CARBON NANOTUBE COATINGS

Geckos possess the remarkable ability to climb vertical walls without using any adhesive glue. A gecko's feet are covered with tiny hairs that stick to rough and smooth surfaces using van der Waals forces. Inspired by the gecko, Ali Dhinojwala and researchers at the University of Akron have developed carbon nanotube–based gecko tapes that stick to most surfaces (even Teflon) and have self-cleaning abilities.

Applying carbon nanotube coatings to steel surfaces makes them superhydrophobic without losing their electrical or thermal conductivity. Modifying the surface characteristics of stainless steel can lead to many new applications such as heat exchangers, electrodes for fuel cells, solar panels, fluid transport, and nonfouling surfaces.

CONTENTS
Carbon nanotubes, steel, polymers

APPLICATIONS
Adhesives, self-cleaning surfaces

TYPES / SIZES
Sheets, films, tubes, rods

ENVIRONMENTAL
Solvent-free systems, allows higher efficiency

TESTS / EXAMINATIONS
Adhesion test, contact angle, and electrical resistance

LIMITATIONS
Cost

CONTACT
The University of Akron
170 University Circle
Akron, OH 44304
www2.uakron.edu/cpspe/dhinojwala/

Emergent Surface

LINEAR ADAPTIVE-SHADING SYSTEM

Based on new technologies for adaptive building skins, Emergent Surface is a wall that continuously reconfigures itself—portions selectively disappear and reappear. In one arrangement, the piece appears as a solid surface with three-dimensional curvature. In another, it resolves itself into seven slender poles, running floor to ceiling. And between these extremes lies an infinite variety of configurations. These different states represent the physical embodiment of digital information. As such, Emergent Surface represents a kind of "material media," operating not on bandwidths of light and sound, but in terms of variable solidity and permeability.

The underlying technology demonstrated by Emergent Surface is Hoberman's linear shading system, composed of a series of individually controlled units that extend from minimal profiles into profiles that can be tailored to custom geometries.

In aggregate, the units can create a computer-controlled cellular shading system. An algorithm combining historic solar gain data with real-time sensing of light levels controls the shading units.

CONTENTS
Machined aluminum struts with steel or acrylic slats

APPLICATIONS
Adaptive coverings over building facades and roofs for commercial, industrial, or residential sectors; suitable for solar shading or rain shedding

TYPES / SIZES
Custom profiles, sizes, and compositions

ENVIRONMENTAL
Controls solar gain and light levels, resulting in decreased energy consumption and improved occupancy comfort; protects from inclement weather and wind

CONTACT
Hoberman Associates
40 Worth Street, Suite 1680
New York, NY 10013
Tel: 212-349-7919
www.hoberman.com
associates@hoberman.com

Expanding Helicoid

ANGULATED SCISSOR LINKAGE

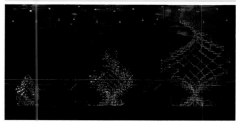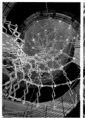

Beginning as a tight cluster, the Expanding Helicoid smoothly expands to fill the spiral staircase of the science museum in which it is housed. Visitors have the remarkable sensation of being inside the sculpture: As it contracts, it seems to disappear into the stairwell; as it expands, it seems to grow like a living plant. Bound by two spirals, like the DNA double helix it resembles, the helicoid itself is like a living organism, evolving as it expands. The form belongs to a class of shapes called minimal surfaces, which occur in nature as soap bubbles and spider webs.

Composed of a series of angulated scissor links, the helicoid is at once both mechanism and structure. Hoberman has used this technology to build a variety of products and installations, including children's toys, deployable shelters, and expanding architecture.

CONTENTS
Machined aluminum hubs and struts

APPLICATIONS
Adaptable to any three-dimensional geometry or surface, such as museum installations, children's toys, or satellites that calibrate radar arrays

TYPES / SIZES
Large range of sizes, materials, and compositions, including a plastic sphere that expands from 2 1/4" to 5" (5.7 cm to 12.7 cm), and an aluminum hyperbolic paraboloid that expands from 15' to 50' (4.5 m to 15.2 m)

ENVIRONMENTAL
Enables deployable buildings for temporary or disaster-relief situations

CONTACT
Hoberman Associates
40 Worth Street, Suite 1680
New York, NY 10013
Tel: 212-349-7919
www.hoberman.com
associates@hoberman.com

Hoberman Arch

RETRACTABLE ARCH

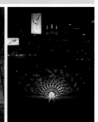

The Hoberman Arch is a seventy-two-foot (twenty-two-meter) diameter transforming "curtain" composed of two main parts: a matrix of movable panels and a static arch that supports those panels. Hoberman's concept of the screen was to use ninety-six panels, each with a skeletal frame constructed from aluminum box sections clad in translucent fiber-reinforced panels. Four differently shaped panels are radially arranged and layered over each other to form an almost solid screen while closed. The outermost panels attach to thirteen radial slides on the static arch. The lower sections of the panels are supported by "trolleys" that ride on tracks housed in the stage's turntable. Thus, all movable elements on the perimeter of the arch are directly attached to static supports. The Hoberman Arch was installed in front of the stage at Olympic Medals Plaza for the 2002 Winter Olympic Games in Salt Lake City. The deployment of the arch was part of a performance that included music, lighting, and dancers. Over five-hundred computer-controlled lights were integrated into the arch's movements so that it entirely changed its color and appearance.

CONTENTS
Steel or aluminum structural supports with glass, plastic, or metal surface cladding

APPLICATIONS
Mechanical curtain or retractable wall to subdivide large spaces for indoor/outdoor use

TYPES / SIZES
72 x 36' (22 x 11 m); 15 tons; custom sizes available

ENVIRONMENTAL
Solar shading and climate control

CONTACT
Hoberman Associates
40 Worth Street, Suite 1680
New York, NY 10013
Tel: 212-349-7919
www.hoberman.com
associates@hoberman.com

ABRASIVE POLISHED STAINLESS-STEEL FINISH

InvariBrush is an abrasive polished finish with a deep, rich, and uniform texture designed for use in architectural applications. Though similar to a standard #4 finish, it is produced with exacting process controls and inspected to a visual standard to ensure the highest degree of visual consistency. Unlike #4, InvariBrush has the uniformity and flatness required for architectural panel applications. InvariBrush can be applied to wall panels, elevator cabs, coping, and trim. Since InvariBrush has no coatings to deteriorate, it will last indefinitely with little maintenance.

CONTENTS
Stainless steel

APPLICATIONS
Wall panels, elevator cabs, coping, trim, various architectural elements

TYPES / SIZES
Thickness .012–.19" (.3–4.8 mm); up to 4' (1.2 m) wide in coils and cut lengths

ENVIRONMENTAL
High recycled content, recyclable

CONTACT
Contrarian Metal Resources
51 QSi Lane
Allison Park, PA 15101
Tel: 724-779-5100
www.metalresources.net
info@metalresources.net

Iris Dome

DOME-BASED RETRACTABLE ROOF

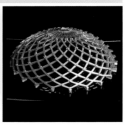

Iris Dome is a retractable roof that opens and closes like the iris of an eye, transforming the space inside smoothly from indoors to outdoors. Metal struts simultaneously carry the load and articulate movement without the assistance of extraneous elements. The fully self-supporting dome can retract completely to the perimeter of an opening, leaving a view unobstructed by support tracks or rigging.

Rigid covering panels are attached to the structure, gliding smoothly over one another to form a continuous skin covering the dome when fully extended. The Iris Dome's flexible design can be customized using parametric modeling tools to adopt a variety of forms and profiles.

CONTENTS
Steel or aluminum structural supports with glass, plastic, or metal surface cladding

APPLICATIONS
Retractable roofs for buildings, arenas, stadiums, pavilions, and plazas

TYPES / SIZES
Design enables a high degree of flexibility; built examples range from a 3' (.9 m) model dome to a 30' (9.1 m) installed dome

ENVIRONMENTAL
Active shading and climate control

CONTACT
Hoberman Associates
40 Worth Street, Suite 1680
New York, NY 10013
Tel: 212-349-7919
www.hoberman.com
associates@hoberman.com

3D SHEET METAL PANELS

Rubber pad–based metal forming is a technique developed by the aerospace industry. Metaalwarenfabriek Phoenix applies this technique to manufacture double-curved sheet metal products for a variety of markets and industries. The advantage of rubber-pad forming is that only one mold is required, since the rubber pad assumes the function of the counter mold and sheet metal holder combined. As a result, tooling costs are only 10–25 percent of those for traditional deep drawing. No-Limit Panels may be used for exterior and interior design applications, and a wide variety of patterns are possible using the rubber-pad forming technique.

CONTENTS

Steel, stainless steel, aluminum, copper, titanium, and other metals or materials that can be formed by molding

APPLICATIONS

Facade coping, wall panels, ceilings, doors, architectural elements

TYPES / SIZES

Maximum size 7.2 x 3.6' (2.2 x 1.1 m); maximum depth/height 5.9" (15 cm); maximum material thickness .12" (3 mm)

ENVIRONMENTAL

100% recyclable

TESTS / EXAMINATIONS

Product simulation in AutoForm

LIMITATIONS

Limitations determined by the characteristics of the chosen material (stretch coefficient)

CONTACT

Metaalwarenfabriek Phoenix BV
Fijenhof 6
Eindhoven, Noord Brabant
5581 AG
The Netherlands
Tel: +31 (0)40-253-29-44
www.phoenixmetaal.com
info@phoenixmetaal.nl

Parabienta

BUILDING AFFORESTATION SYSTEM

Parabienta is a vertical greening system for buildings developed by Japan-based Shimizu Corporation. Composed of units that incorporate Excelsoil solidified soil base in stainless-steel wire frames, a lush green wall may be achieved quickly with the installation of the system. Parabienta has been shown to reduce solar heat gain on facades, thus reducing building energy costs. The system mitigates noise transfer as well, especially in the high-frequency range. Parabienta vertical greening units can be arranged to make various designs. Different kinds of plants can be utilized for different colors and textures as well as different performance criteria. In addition, units may easily be relocated or replaced when design or maintenance needs dictate.

CONTENTS
Stainless-steel wire, stainless-steel channel, solidified soil

APPLICATIONS
Green wall, garden

TYPES / SIZES
Standard unit size 22.1 x 22.1" (56 x 56 cm)

ENVIRONMENTAL
Temperature reduction of wall (mitigating the heat island phenomenon, reducing the thermal load for air-conditioning); noise reduction

LIMITATIONS
Only available in Japan

CONTACT
Shimizu Corporation
3-4-17 Etchujima
Tokyo, Koyo-ku 135-8530
Japan
Tel: +81 3-3820-5504
www.shimz.co.jp/english/index.html
www.shimz.co.jp/toiawase/contact.html

Photo-Engraved Aluminum

ALUMINUM PANEL WITH PERMANENT PHOTOGRAPHIC TRANSFER

The focus of Intaglio Composites is the art of photographic transfer to materials. The company has developed a new, recyclable process that allows the permanent engraving of images, graphics, and text into aluminum. Unlike laser engraving, there is no restriction on size, the etching depth may be varied, and the process is relatively inexpensive. Photo-Engraved Aluminum accepts either a clear-anodized or an industrial clear-coat finish, and a wide range of colors can be achieved through the application of patinas. It is wear- and graffiti- resistant for use in interior and exterior environments, and may be applied to flat or curved surfaces.

CONTENTS
100% aluminum

APPLICATIONS
Wall panels, architectural signage, public art

TYPES / SIZES
Engraving types: halftone images, vector graphics; shapes: flat or convex, concave, and compound-curve panels (custom shapes possible)

ENVIRONMENTAL
100% recyclable material; 85% recyclable engraving process

CONTACT
Intaglio Composites
3101 Pleasant Valley Lane
Arlington, TX 76015
Tel: 817-465-2773
www.intagliocomposites
.com
sales@intagliocomposites
.com

Reynobond with Kevlar

KEVLAR-COATED ALUMINUM COMPOSITE PANEL

The devastating power of hurricanes has had a memorable impact in recent years, and damage caused by wind-borne debris has destroyed many commercial structures. Alcoa has developed a new hurricane-resistant panel made with DuPont Kevlar—a material five times as strong as steel by weight. Reynobond with Kevlar adds a thin layer of Kevlar fabric to a polyethylene core, making a lightweight panel that does not require heavy backer materials.

Reynobond with Kevlar can achieve complex design specifications such as angles and curves, while offering the flat, smooth, and consistent surface that aluminum composite panels are known for. Available in standard Kynar/PVDF resin–based paint systems with twenty-year performance warranties, maintenance costs are lower than for concrete, brick, or stucco. Reynobond with Kevlar installed with Alcoa-designed extrusions is the first aluminum composite panel system to eliminate the need for protective backer materials such as plywood, steel, or concrete. Panel modules can be shop-fabricated and quickly installed onto structural steel studs on the job site. This process not only decreases the material cost of installation, but also reduces on-site labor requirements, leading to a faster installation with costs comparable to—and often less than—those of brick, exterior insulation finishing systems (EIFS), or stucco.

CONTENTS
Extruded polyethylene core and Kevlar fabric

TYPES / SIZES
Widths 62" (157.5 cm), 50" (127 cm); lengths 48–243" (122–617 cm); thickness .16" (4 mm)

TESTS / EXAMINATIONS
Meets Miami-Dade Building Code–based small and large missile impact tests

LIMITATIONS
10 panel minimum purchase in conjunction with a production run of same color and width

CONTACT
Alcoa Architectural Products
50 Industrial Boulevard
Eastman, GA 31023
Tel: 800-841-7774
www.reynobond.com

ULTRALIGHT GRAPHITE POLYMER COMPOSITES

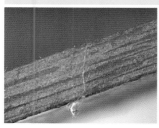  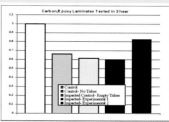

TRANSFORMATIONAL PRODUCT

Aircraft and marine vessels are significant contributors to global warming. Airplane vapor trails release CO_2 as well as other pollutants, and ships expend large amounts of energy as they carry most of the world's cargo. Many of the polymer composites typically used for such craft are over-engineered to avoid structural failure, a fact that reduces the advantage of lightweight constructions.

Self-Repairing Composites—created by Carolyn Dry—are made from graphite oil, resulting in lighter material properties. In a recent project for the U.S. Air Force, the composite portion of the airplane fuselage was made to be 30 percent lighter than when made with conventional materials, based on the use of Self-Repairing Composites. If this material can be successfully implemented, a significant reduction in CO_2 contributions from traveling craft and other applications will be realized.

CONTENTS
Graphite

APPLICATIONS
Airplanes, ships, infrastructure

ENVIRONMENTAL
Fuel savings

TESTS / EXAMINATIONS
Complete set of ASTM tests of flexure, compression, compression after impact, shear

LIMITATIONS
Polymer use relies on petroleum and graphite is sometimes difficult to obtain

CONTACT
Natural Process Design Inc.
1250 East 8th Street
Winona, MN 55987
www.naturalprocessdesign
.com
naturalprocessdesign@
yahoo.com

SolPix

SOLAR-POWERED SUN-SHADING MEDIA WALL

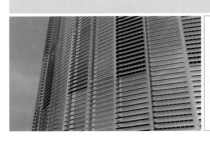

SolPix is a patented solar-powered media wall for medium- to large-scale installations in new construction or existing buildings. Developed by Simone Giostra, SolPix is a completely integrated system for power production and sun shading, and also acts as a digital screen. SolPix allows for dynamic content display, including playback videos, interactive performances, and live- and user-generated content. The "intelligent skin" interacts with building interiors and external public spaces using embedded, custom-designed software, transforming a building facade into a responsive environment for entertainment and public engagement.

The panels can be used to create stunning media effects on very large building envelopes that are viewable from both inside and outside a building. The photovoltaic system does not need to be in direct sunlight to work, and will generate electricity even on cloudy days. The panels have a power warranty of twenty years and are expected to generate power for fifty years.

SolPix allows daylight into the building while reducing its exposure to direct sunlight. The sun-shading elements provide unobstructed outside views from the building interior, while lending a contemporary texture to the building exterior. The horizontal or vertical panels can be mounted at a preferred angle or can be rotated in order to maximize exposure to direct sunlight.

CONTENTS
Medium-resolution LED lighting, photovoltaic cells, extruded aluminum

APPLICATIONS
Solar-powered media wall for medium- to large-scale installations in new construction or existing building envelopes, freestanding billboards, large environmental graphics, ambient media installations

TYPES / SIZES
Variable resolution, no size limitation

ENVIRONMENTAL
Renewable energy source, solar gain control

TESTS / EXAMINATIONS
UL-Certified

LIMITATIONS
Requires exposure to direct or indirect sunlight

CONTACT
SolPix LLC
55 Washington Street, #454
Brooklyn, NY 11201
Tel: 212-920-8180
www.solpix.org
info@solpix.org

ARCHITECTURAL MESH SHADING SYSTEM

Solucent is an energy-saving, daylighting mesh-shading system for building exteriors and interiors. It was developed by Cambridge Architectural to meet the ever-increasing energy-saving needs facing architects today, but without sacrificing beauty in design. The system combines the aesthetic qualities of architectural mesh with shading and daylighting capabilities to create a compelling energy- and light-management solution. Solucent mesh shading systems reduce solar heat gain, leading to significant savings on cooling costs. Solucent is also a versatile and transparent daylighting material that can allow the desired amount of natural light into a building without obstructing views out of the windows. As a result, Solucent systems seamlessly reduce the need for electric light, which is the number one energy consumer in buildings and a contributor of unwanted heat gain. Each Solucent system is designed based on a building's solar orientation. The mesh pattern is also chosen according to these specifications, allowing each piece of the system to fit together precisely and effectively.

CONTENTS
Stainless steel

APPLICATIONS
Solar heat-gain reduction, daylighting, ventilation, maintaining views

TYPES / SIZES
Custom-built; panels of up to 100' (30.5 m); can be installed in tension on building exteriors

ENVIRONMENTAL
Readily recyclable and manufactured from a high percentage of recycled materials, reduces building cooling costs, reduces the need for electric lighting

TESTS / EXAMINATIONS
Solar heat-gain reduction reports, headlight attenuation reports, Miami-Dade Certification

CONTACT
Cambridge Architectural
105 Goodwill Road
Cambridge, MD 21613
Tel: 866-806-2385
www.cambridgearchitectural.com
sales@cambridgearchitectural.com

Vault-Structured Metal

ULTRALIGHT PATTERNED METAL PANELS

Numerous phenomena can be observed in nature that are the result of controlled self-organization. Bionic vault-structuring is a method to generate a three-dimensional pattern in metallic sheets. Thanks to the controlled self-organization arrangement, a minimum of plastic deformation is required for forming the patterns. Because of their high rigidity, vault-structured, hexagonally or three-dimensional facet-structured components can be produced in stainless steel with greatly reduced wall thickness.

Vault-structured materials, even if thin and lightweight, are highly resistant to bending and to stress caused by thermal expansion, and they have other advantageous properties with application potential for lightweight structures. When flow bypasses three-dimensional profiled surfaces, there is a higher convective heat and mass-transfer coefficient compared to smooth surfaces.

CONTENTS
Steel, stainless steel, aluminum, copper, or brass (also paper or cardboard)

APPLICATIONS
Lightweight products for exterior and interior applications

TYPES / SIZES
Hexagonal or three-dimensional facette-structured pattern; hexagon (wrench size) .6", 1.3", 1.5", and 2" (1.6 cm, 3.3 cm, 3.9 cm, and 5.1 cm); width of sheet metal or coils up to .04" (1.1 mm)

ENVIRONMENTAL
Reduction of weight/wall thickness roughly 30% compared with flat sheet metal; reduction of process energy for manufacturing of three-dimensional-structured (high-stiffened) sheet metal

TESTS / EXAMINATIONS
Bending stiffness; flexural strength

LIMITATIONS
Maximum wall thickness of vault-structured sheet metal: .04" (1 mm) steel, copper; .03" (.8 mm) stainless steel; .05" (1.2 mm) aluminum

CONTACT
Dr. Frank Mirtsch
Ruhlsdorfer Strasse 95,
Gebäude 101
Stahnsdorf, Berlin/
Brandenburg 14532
Germany
Tel: 0049 3329-699-576
www.mirtschusa.com
saboorian@mirtschusa.com

04: **WOOD**

3D Veneer

THREE-DIMENSIONAL SHAPABLE-WOOD VENEER

3D Veneer is a three-dimensionally formable veneer for the industrial production of plywood moldings. Developed by Reholz GmbH, the process allows conventional wood veneers to be treated in such a way that they can be "deep drawn," preserving the texture of the wood. Several layers of 3D Veneer can be glued to a plywood shape, and conventional plywood production machinery may then be used to modify the veneer.

In addition to the fabrication of molded parts, 3D Veneers may be also used for coating three-dimensional components in a wide range of materials. A recent enhancement includes a process called FFT (form following texture), in which the veneer grain follows a variable curve.

CONTENTS

100% wood (veneer texture is completely preserved)

APPLICATIONS

Plywood Shapes: seating shells, backrests, automotive industry moldings, coating of plastic housings, cabinet doors, musical instruments, sports equipment, yacht or caravan interiors; As FFT: tables, edge covers, and interiors. 3D Veneer surfaces can be stained and lacquered.

TYPES / SIZES

Beech, oak, maple, walnut, or Vinterio; maximum size 78.7 x 38.6" (200 x 98 cm); can be joined to form a larger width; thickness .05" (1.15 mm)

ENVIRONMENTAL

Can reduce the weight of moldings if the shape is three-dimensionally formed, due to high stiffness

TESTS / EXAMINATIONS

Chair shells made with five layers 3D Veneer were successfully tested according to standard EN 1728 and DIN V ENV 12520

LIMITATIONS

Interior use only (veneers for outdoor use are currently under development)

CONTACT

Reholz GmbH
Sachsenallee 11
Kesselsdorf, Saxony 01723
Germany
Tel: +49 35204-780430
www.reholz.de
info@reholz.de

BIO-BASED STRUCTURAL INSULATED PANEL

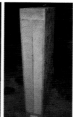

Agriboard is a company that makes insulated building panels comprised of compressed agricultural fibers, oriented strand board, and engineered lumber caps. Its panels are made with agricultural fibers, a rapidly renewable, bio-based, sustainable and readily-recyclable resource. Since no adhesives are added to the panels, there are no unwanted odors or emissions. The panels offer a dynamic insulation value in excess of R-25, and may be easily machined to match design requirements. Agriboard panels are also fungi- and pest-resistant because they hold no nutritional value.

CONTENTS
Compressed agricultural fiber, engineered lumber, oriented strand board

APPLICATIONS
Commercial or residential buildings up to three stories

TYPES / SIZES
Maximum length 24' (7.3 m); width 4' (1.2 m); thickness 3 1/2" (8.9 cm)

ENVIRONMENTAL
Job-site waste minimization, low carbon footprint

TESTS / EXAMINATIONS
ICC approved, FEMA approved

CONTACT
Agriboard Industries
100 Industrial Park Drive
Electra, TX 76360
Tel: 940-495-3590
www.agriboard.com
sales@agriboard.com

BAMBOO

3D BAMBOO PANELS

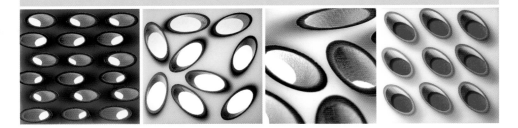

BAMBOO panels are made by casting bamboo sections in translucent resins. The diagonally cut bamboo sections may be left open or filled with a different color resin for a dramatic spatial effect. By changing the amount, size, or direction of the bamboo ovals, the acoustic performance of the panels changes, as well as their look and feel. Their simple production technique allows the panels to be easily customized based on aesthetic and functional requirements.

Designers Yvonne Laurysen and Erik Mantel developed BAMBOO for the exhibit Design Interventions for Stimulating Bamboo Commercialization by Pablo van der Lught.

CONTENTS
Bamboo, resin

APPLICATIONS
Wall cladding, space divider, variable-acoustic noise reduction, aesthetic applications

TYPES / SIZES
23.6 x 23.6" (60 x 60 cm); maximum thickness 1" (2.5 cm); custom sizes and colors possible

ENVIRONMENTAL
Rapidly renewable materials

LIMITATIONS
Not for exterior use

CONTACT
LAMA Concept
Elektronstraat 12, #5
Amsterdam, Noord-Holland
1014 AP
The Netherlands
Tel: +31 20-4121-798
www.lamaconcept.nl
info@lamaconcept.nl

CROSS-LAMINATED TIMBER PANELS

BBS are structural solid-wood panels made of sustainably harvested, fast-growth softwood. By laminating longitudinal and cross layers (X-LAM), the natural forces of the wood (expanding, contracting, bending) are minimized to achieve great strength and stability. The panels are used primarily for structure, but can also provide a finished, exposed surface. The BBS panels are produced through a digital fabrication process where the shop drawings are transmitted to the plant as three-dimensional CAD files. The panels are then CNC-cut and shipped directly to the construction site.

CONTENTS
99.4% solid timber, 0.6% formaldehyde-free glue

APPLICATIONS
Self-supporting structural floors and roofs, load-bearing walls

TYPES / SIZES
Spruce, larch, Douglas fir, pine, white fir; width 4.1' (1.25 m); maximum length 78.7' (24 m)

ENVIRONMENTAL
Carbon neutral; sustainably harvested wood (PEFC certified)

TESTS / EXAMINATIONS
European Technical Approval ETA-06/0009, German Technical Approval Z-9.1-534

LIMITATIONS
For shipping to the United States and Canada, maximum panel length is 38' (11.6 m)

CONTACT
HolzBuild
19 Country Club Lane
Briarcliff Manor, NY 10510
Tel: 914-908-5238
www.holzbuild.com
ag@holzbuild.com

Coco Tiles

ED COCONUT SHELL TILES

Made from reclaimed coconut shells, Kirei Coco Tiles may be used both horizontally and vertically as decorative tiles or panels. Featuring multiple pattern and color combinations and available in light, dark, and mixed textures, the coconut-shell tiles create a variety of surfaces and enhance the sustainable material palette in residential, commercial, and hospitality applications.

CONTENTS
Coconut shells, low-VOC resin, sustainably harvested wood backer mat

APPLICATIONS
Architectural millwork, interior design, wall coverings, cabinetry, retail displays, hospitality, furniture, restaurants, finished products, hotels, commercial, residential

TYPES / SIZES
Tiles 11.8 x 11.8" (30 x 30 cm); panels 47.2 x 47.2" (120 x 120 cm); backer thickness .4" (9 mm)

ENVIRONMENTAL
Reduces landfill waste and air pollution; the shells are a rapidly renewable resource left after the edible portion of the coconut is harvested

TESTS / EXAMINATIONS
Fire Class C

LIMITATIONS
Not recommended for exterior use; tiles should be conditioned for a minimum of 72 hours before installation

CONTACT
Kirei USA
412 North Cedros Avenue
Solana Beach, CA 92075
Tel: 619-236-9924
www.kireiusa.com
info@kireiusa.com

SUSTAINABLE HARDWOOD PANELS

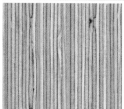   

Ecolinea wood paneling is a new product to the North American market, initially developed for the window industry. With its distinctive linear grain, Ecolinea is composed of a combination of Douglas fir harvested from sustainably managed forests and MDF backing. The material is suitable for millwork applications and feature walls, and may be visually coordinated with ASI's Linea Nova S/V/L flooring.

CONTENTS
Douglas fir, medium-density fiberboard (MDF) backing

APPLICATIONS
All millwork applications, including store fixtures, furniture, and feature walls

TYPES / SIZES
4 x 8' (1.2 x 2.4 m) or 4 x 10' (1.2 x 3 m); thicknesses .75" (1.9 cm), 1.18" (3 cm), 1.65" (4.2 cm); veneer .08" (2 mm) laminated

ENVIRONMENTAL
Douglas fir harvested under the strict guidelines of the Canadian Standards Association, which operates in conjunction with the Sustainable Forestry Initiative; formaldehyde-free panels available; MDF made from recycled content

LIMITATIONS
Interior use only

CONTACT
Architectural Systems, Inc.
150 West 25th Street,
8th Floor
New York, NY 10001
Tel: 800-793-0224
www.archsystems.com
sales@archsystems.com

Kebony

KEBONIZED WOOD BASED ON PINE, MAPLE, AND BEECH

Kebony is a high-performance wood that is modified by a process called Kebonization, which is an environmentally friendly procedure that enhances the properties of wood using biowaste from the sugar industry. Kebony is a durable alternative to impregnated surface-treated and tropical timber. The process, which is based on a liquid extracted from biowaste, strengthens the cellular walls of wood, increases the density of the materials, and makes the product stiffer and significantly harder than untreated wood. Kebonization results in the wood cells being permanently blocked, which reduces shrinkage and swelling by approximately 50 percent when compared with untreated wood. The polymer is permanently bonded to the cell structure in the wood by means of a process that cannot be reversed; thus, Kebony contains no chemicals that can be released into the environment. In the waste disposal phase, Kebony can be treated as regular untreated wood.

Available Kebony species are pine, spruce, oak, beech, maple, and southern yellow pine. The raw materials for Kebony are acquired from commercially managed forests with large timber harvests. Kebonized wood has a golden brown color that naturally acquires a silvery gray patina, and exposure to sun and rain creates an interesting effect of visual depth. Kebony exhibits good durability and long life spans in harsh climates, and there is no need for paint or sealing. The increased resistance protects against decay, fungi, insects, and other microorganisms. Required maintenance is limited to normal cleaning.

CONTENTS
Wood, biowaste from the sugar industry

APPLICATIONS
Cladding, decking, roofing, construction, boating, furniture, floors

TYPES / SIZES
Large number of standardized profiles and dimensions

ENVIRONMENTAL
Reuse of biowaste, sourced from commercially managed forests, long life spans, no treatment needed, nontoxic (exempted from EU's biocide directive 76/769/EEC)

TESTS / EXAMINATIONS
On demand

LIMITATIONS
Fasteners must be stainless steel (or acid-proof)

CONTACT
Kebony ASA
Havneveien 35
Skien, N-3739
Norway
Tel: +47 06125
www.kebony.com
info@kebony.com

Kirei Bamboo

BAMBOO PANELS

Kirei Bamboo is an eco-friendly modern millwork material manufactured from the fast-growing trunks of the Moso bamboo grass with a low- or no-added-urea formaldehyde adhesive. Bamboo is a rapidly renewable resource with a fast growth cycle, resulting in higher material yield per acre than tree planting. Kirei Bamboo paneling is strong and dense, and can be used in a wide variety of surface and millwork applications.

CONTENTS

Rapidly renewable Moso bamboo, low- or no-added-urea formaldehyde adhesive

APPLICATIONS

Architectural millwork, interior design, wall covering, cabinetry, retail displays, flooring, hospitality, furniture, restaurants, finished products, hotels, commercial, residential

TYPES / SIZES

Chocolate, Horizontal Natural, Vertical Natural, Horizontal Carbonized, Vertical Carbonized, Zebra 1, Zebra 2, Horizontal Zebra

ENVIRONMENTAL

Rapidly renewable, low VOCs

CONTACT

Kirei USA
412 North Cedros Avenue
Solana Beach, CA 92075
Tel: 619-236-9924
www.kireiusa.com
info@kireiusa.com

Kirei WheatBoard

ECO-FRIENDLY MDF ALTERNATIVE

Kirei WheatBoard was developed as an environmentally sensitive alternative to formaldehyde-emitting medium-density fiberboard (MDF) products, with working characteristics meeting or surpassing those of commercially available MDF. Kirei WheatBoard is made with reclaimed wheat stalks and non toxic adhesives, and may be used in millwork, cabinetry, and finished-wood product applications. Kirei WheatBoard can be painted or laminated with a wide variety of surface treatments, including Kirei Bamboo Veneers.

CONTENTS
Reclaimed wheat stalks, no-added-formaldehyde MDI adhesive

APPLICATIONS
Architectural millwork, interior design, wall covering, cabinetry, retail displays, flooring, hospitality, furniture, restaurants, finished products, hotels, commercial, residential

TYPES / SIZES
4 x 8' (1.2 x 2.4 m); standard thickness 1/2" (1.3 cm) or 3/4" (1.9 cm)

ENVIRONMENTAL
Reduces forest clear-cutting, renewable resource, low VOCs

TESTS / EXAMINATIONS
Class C (Class A-rated door core available, WheatBoard may achieve Class A using flame retardants)

LIMITATIONS
Not recommended for exterior use

CONTACT
Kirei USA
412 North Cedros Avenue
Solana Beach, CA 92075
Tel: 619-236-9924
www.kireiusa.com
info@kireiusa.com

Kraftplex

WOOD FIBER SHEETING

Kraftplex combines the characteristics of metal, composite materials, and plastics. Composed of 100 percent unbleached wood fibers with no added binders or adhesives, it is a stable, flexible, and durably shapeable material akin to metal sheets. Because Kraftplex has electrically isolating properties, it is a good alternative to conventional plastic sheeting. It is also flexible enough that three-dimensional shaping via deep-drawing and edge-shaping are possible. Kraftplex may be easily cut, drilled, perforated, and adhered using common woodworking tools. It can also be treated with paints, varnishes, oils, waxes, and adhesives, as well as covered with decor film and printed.

CONTENTS
100% unbleached wood fibers, no added adhesives or binders

APPLICATIONS
Furniture, tool manufacturing, electronic components, product design, interior design, industrial design

TYPES / SIZES
Sheet size 96 x 55" (244 cm x 140 cm); thickness .03" (.76 mm) or .06" (1.5 mm)

ENVIRONMENTAL
100% recyclable; sustainably cultivated softwood stocks; no chemical additives, bleaches, or binding agents used

TESTS / EXAMINATIONS
UL94HB

LIMITATIONS
Not for exterior use, avoid humid environments

CONTACT
Well Ausstellungssystem GmbH
Schwarzer Bär 2
Hannover, D-30449
Germany
Tel: +49 511-92881-10
www.well.de
info@well.de

Laser-Cut Cork

LASER-CUT CORK COMPOSITE SURFACES

Laser-Cut Cork surfaces are made from the by-product of wine stopper production. Developed by Yemi Awosile, the material benefits from the natural acoustic and thermal insulating properties found in cork, and is well suited to interior design applications such as wall panels, tapestries, and room dividers. Laser-Cut Cork surfaces are available in a raw cork composite grain, glossy black, or various metallic finishes such as silver, copper, or gold. The material is also available in a variety of designs and can be custom made to suit particular needs.

CONTENTS
Cork composite

APPLICATIONS
Wall covering, acoustic wall panels

TYPES / SIZES
Available in a variety of custom sizes

ENVIRONMENTAL
Thermal insulating properties

TESTS / EXAMINATIONS
Pending

LIMITATIONS
Not for external use

CONTACT
Yemi Awosile
66B Elgin Crescent
London, W11 2JJ
United Kingdom
www.yemiawosile.co.uk
yemi.awosile@network.rca
.ac.uk

LINEAR-GRAIN HARDWOOD FLOORING

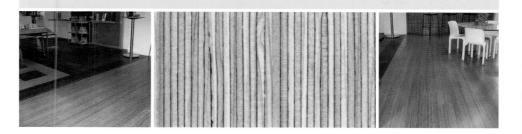

Linea Nova is a new engineered-hardwood flooring made from sustainably harvested timber. Its distinctive linear wood grain coordinates with Architectural System, Inc.'s Ecolinea wood panels, which are also made from Douglas fir. Linea Nova has tongue-and-groove construction for ease of installation, and features a two-ply, UV-cured acrylic topcoat finish.

CONTENTS
Douglas fir

APPLICATIONS
Flooring for all commercial environments

TYPES / SIZES
96 x 7 1/4 x 1/2" (244 x 18.5 x 1.2 cm)

ENVIRONMENTAL
Harvested under the strict guidelines of the CSA (Canadian Standards Association), which operates in conjunction with the SFI (Sustainable Forestry Initiative)

LIMITATIONS
Interior use only

CONTACT
Architectural Systems, Inc.
150 West 25th Street,
8th Floor
New York, NY 10001
Tel: 800-793-0224
www.archsystems.com
sales@archsystems.com

Made to Measure

ADJUSTABLE STOOL

Frustrated by the number of times a stool didn't quite suit the height of the surface it was selected for, Monika Piatkowski designed the Made to Measure stool—a customizable solution with built-in saw and ruled legs. A tongue-in-cheek example of consumer-adaptable design, the Made to Measure stool is made from solid hardwood legs with a wide range of ruled lines. The seat is made from veneered MDF, layered in two sections to make room for a build-in metal saw with a wooden handle.

CONTENTS
Stained hardwood legs, veneered medium-density fiberboard (MDF) seat, built-in metal saw

APPLICATIONS
Seating

TYPES / SIZES
12.6 x 12.6 x 27.2" (32 x 32 x 69 cm)

LIMITATIONS
Interior use only

CONTACT
Hive
The Studio,
6 Shardeloes Road
London, SE14 6NZ
United Kingdom
Tel: +44 (0)20-8692-0219
www.hivespace.com
hive@hivespace.com

MYCOLOGICAL BIOCOMPOSITE

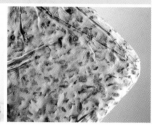

Mycobond, developed by Edward Browka, is a mycological biocomposite that can be used in a wide variety of applications. Instead of conventional manufacturing processes, Mycobond uses mycelium—which is essentially the root system of a mushroom—to transform loose aggregates into strong composites. This process can be varied by using different species of fungus and mixtures of aggregates in order to make a composite with an optimal density, strength, appearance, and performance for the specific application.

Additionally, Mycobond represents a low-embodied-energy manufacturing process as the material self-assembles at room temperature in the dark. Furthermore, Mycobond upcycles resources like rice hulls, cotton burrs, and buckwheat hulls that are otherwise thrown away, transforming them into valuable products, including rigid board insulation and protective packaging buffers.

CONTENTS
Agricultural waste products, mycelium

APPLICATIONS
Rigid board insulation, protective packaging buffers, acoustic panels, surfboard cores

TYPES / SIZES
Size varies according to application

ENVIRONMENTAL
100% natural, biodegradable, low-embodied-energy manufacturing process, locally produced, uses renewable and neglected resources from local waste streams

TESTS / EXAMINATIONS
ASTM C165-07, C203-05, D3500-90, C1338, C1134, C177; Class A Fire Rating, Cone Calorimeter ASTM E 1354

CONTACT
Ecovative Design
1223 Peoples Avenue
Troy, NY 12180
Tel: 518-690-0399
www.ecovativedesign.com
info@ecovativedesign.com

Placage

HIGH-CONTRAST LINEAR HARDWOOD VENEER PANELS

Placage is a bespoke wood veneer collection, available in five striped patterns, from subtle to vibrant, in varying widths. This ASI Wood Panel product offers several striking design choices in two panel sizes, and is perfect for all millwork applications including store fixtures, furniture, and feature walls.

CONTENTS
Wood veneer, medium-density fiberboard (MDF) or Placorex backing

APPLICATIONS
Store fixtures, furniture, and feature walls

TYPES / SIZES
98 x 48" (249 x 122 cm) on a .63" (16 mm) MDF core; 85 x 37" (216 x 94 cm) on .06" (1.5 mm) Placorex (back)

ENVIRONMENTAL
Programme for the Endorsement of Forest Certification (PEFC) certified material

LIMITATIONS
Interior use only

CONTACT
Architectural Systems, Inc.
150 West 25th Street,
8th Floor
New York, NY 10001
Tel: 800-793-0224
www.archsystems.com
sales@archsystems.com

Wood

EXTENDABLE BOOKCASE

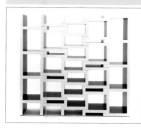 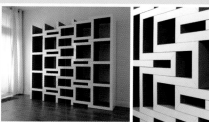 

TRANSFORMATIONAL PRODUCT

REK is a bookcase that grows with one's book collection. The more books, the bigger the bookcase. The zigzag-shaped, 2.4-inch (6-centimeter) thick parts slide in and out, providing as much space as needed. In this way, REK will always be full regardless of the quantity of books. In addition, the varying spaces that appear allow one to arrange books according to their size.

REK is constructed with a high level of precision. In the standard configuration, the outside of REK is finished with white high-gloss laminate, and the inside is finished with warm-gray satin laminate. Custom laminate colors are also available.

CONTENTS
Wood, plastic laminate

APPLICATIONS
Shelving

TYPES / SIZES
Expandable; shelf thickness
2.4" (6 cm); custom laminate
colors available

CONTACT
Reinier de Jong
Coolhaven-terras, 2d
Rotterdam, 3024 AT
The Netherlands
Tel: +31 (0)645724852
www.reinierdejong.com
info@reinierdejong.com

Shelf Space

CNC-MACHINED HARDWOOD

The result of an experimental four-month collaboration with an aerospace machinery manufacturer, the fluid form of Paul Loebach's Shelf Space pushes the limits of wood engineering and advanced machining technology. There have been vast developments in the evolution of CNC-machining technology in the last thirty years, and this product applies sophisticated modern manufacturing techniques to a traditional renewable material.

Shelf Space is made from a stack lamination of solid wood, cut into shape using a multiaxis milling machine normally used for machining aerospace parts. The decorative form is inspired by the language of eighteenth-century woodworking and the shape—nearly impossible to create with conventional tools—is designed to broaden one's expectations of what can be called "traditional."

Three-dimensional computer modeling facilitated the design of a precise yet fluid form. Although it took months to perfect the programming of the complex tool path, due to the incredible power of the machinery each shelf can now be made in just twenty minutes.

CONTENTS
Basswood

APPLICATIONS
Indoor furniture

TYPES / SIZES
45 x 15 x 21" (114 x 38 x 53 cm)

ENVIRONMENTAL
100% sustainably harvested, renewable resource

LIMITATIONS
For indoor use only

CONTACT
Paul Loebach
144 Spencer Street, #608
Brooklyn, NY 11205
Tel: 646-489-2749
www.paulloebach.com
paul@paulloebach.com

Songwood

COMPRESSED-WOOD COMPOSITE

Engineered Timber Resources focuses on the use of by-product wood fibers in the production of decorative wood products, with the intent to limit deforestation. Songwood is a 100 percent recycled, FSC-certified raw material made from by-products from the furniture-manufacturing and pulp industries. The wood fibers are kiln-dried, mixed with a low-VOC resin, and then compressed under eighteen-hundred tons of pressure before being cured for stability. The resulting log can then be used in the manufacturing of any wood-based product, both for interiors and exteriors.

Songwood is not only very durable and hard-wearing, but also extremely stable compared with traditional wood products. In addition to the functional benefits of the product, there are also numerous aesthetic characteristics that are customizable based on the raw material inputs and the desired outcome. Songwood can be infused with organic dyes and colorants prior to the compression, with the net result being a solid-body, integral color, that is therefore sandable.

CONTENTS
100% postindustrial wood fibers

APPLICATIONS
Flooring, paneling, general wood products

TYPES / SIZES
Custom

ENVIRONMENTAL
100% reclaimed industrial by-product; low-VOC; FSC "pure" label; urea formaldehyde-free; low life-cycle costs

TESTS / EXAMINATIONS
ASTM E-84 (Class I for flame spread); formaldehyde (E-1 compliant); MSDS

LIMITATIONS
Exterior application requires the use of a different glue

CONTACT
Engineered Timber Resources
1900 55th Street, #104
Boulder, CO 80301
Tel: 303-440-8842
www.etimberr.com
info@etimberr.com

Structural Tambour

WOOD-FABRIC HYBRID TAMBOUR

Designer Hongtao Zhou invented Structural Tambour in 2008 in Madison, Wisconsin, and made his first bench from this new structure. The idea of Structural Tambour came from decorative Chinese bamboo strips. Dr. Zhou developed the idea into an engineered structure with strong fabric stapled or bonded on the back of wood strips. In this way, Structural Tambour is given flexibility as well as strength so that it can be used for functional furniture design, feature walls, sculpture installations, and temporary footbridges. The advantage of the structure is that it can be made of scrap wood or small-diameter wood to form a larger structural curved surface with both flexibility and stability. It can also be rolled up for storage or transportation after usage.

CONTENTS
Wood, fabric, staples

APPLICATIONS
Furniture, wall features, sculptural installations, flooring

TYPES / SIZES
Custom

ENVIRONMENTAL
Lightweight construction, recyclable materials

CONTACT
University of Wisconsin-Madison
455 North Park Street
Madison, WI 53706
Tel: 608-658-0723
www.uwwood.com/uw_wood/_program.html
lifeisfurniture@gmail.com

HARDWOOD-VENEER FLOORING WITH CORK SUBSTRATE

RECOMBINANT MATERIAL

VenCork combines the refined aesthetic qualities of natural wood with the resilience of a cork substrate. This durable hybrid comes in 4-inch (10.2-centimeter) or 6-inch (15.2-centimeter) wide planks and in eight colors, with the option for custom treatments. VenCork also has a durable, low-VOC, transparent antiscratch wear layer for enhanced longevity.

CONTENTS
Wood veneer, cork substrate

APPLICATIONS
Flooring

TYPES / SIZES
Standard plank thickness 1/8" (3.2 mm); widths 4" (10.2 cm) or 6" (15.2 cm); length 35 7/16" (90 cm)

ENVIRONMENTAL
Recycled content, rapidly renewable materials, low VOCs (GREENGUARD certified)

LIMITATIONS
For interior use only

CONTACT
Architectural Systems, Inc.
150 West 25th Street,
8th Floor
New York, NY 10001
Tel: 800-793-0224
www.archsystems.com
sales@archsystems.com

Wave Wall

CURVILINEAR WALL AND CEILING PANEL SYSTEM

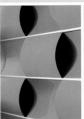

Wave Wall is an undulating, curvilinear wall constructed of refined panels available in either continuous or alternating patterns. Wave Wall is a modular system that can be applied horizontally or vertically, and is offered in three module lengths. The system is easy to install and is available in a wide range of tile finishes. Lighting from behind, front, and above can create both subtle and bold effects, depending on the intensity of the lighting and the choice of tile finishes.

CONTENTS
Timber laminate, timber veneer, polypropylene, plastic laminate, metal laminate, aluminum or printed polyethylene terephthalate (PETG) tiles, aluminum frame

APPLICATIONS
Interior wall and ceiling features—can be combined with backlighting for effect; vertical or horizontal orientation

TYPES / SIZES
Modular units 10.6 x 19.5/25.5/39.4"
(27 x 49.5/75/100 cm)

ENVIRONMENTAL
Rapidly renewable, recycled, and recyclable content; easily disassembled and reassembled

TESTS / EXAMINATIONS
Available upon request

CONTACT
Wovin Wall
6-8 Ricketty Street
Mascot, NSW 2020
Australia
Tel: +61 2-9317-0222
www.wovinwall.com
esales@wovinwall.com

05: **PLASTIC + RUBBER**

3S Solar Block

ENERGY-EFFICIENT DIFFUSE DAYLIGHTING SYSTEM

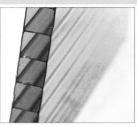 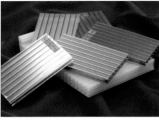 

Duo-Gard Daylighting with 3S Solar Block coating represents an advance in high-performance daylighting technology using translucent polycarbonate glazing. The vision-enhanced, spectrally selective coating is applied to the interior of the sheet's structural cells in order to block all ultraviolet light and 90 percent of infrared light, meanwhile allowing up to 70 percent visible-light transmission. The results are lower energy costs, improved interior environments, and higher productivity.

3S Solar Block brings higher performance to all applications requiring the benefits of natural light without heat gain. It reduces a building's heat load by up to 50 percent, depending on the coating's thickness—allowing for the installation of smaller, more efficient air-conditioning equipment that enjoys longer life spans and less maintenance.

The coating, framing, and shatter-resistant, self-extinguishing polycarbonate glazing are all available in many complementary colors.

CONTENTS
Multiwall translucent polycarbonate sheeting, spectrally selective coating, aluminum frame

APPLICATIONS
Skylights, clerestories, windows, curtain walls, canopies, overglazing, shading systems, bus shelters, walkways; appropriate where high diffuse natural light and high energy efficiency are desired

TYPES / SIZES
Custom design/build systems integrating glazing, coating, and framing

ENVIRONMENTAL
Has exceptional energy efficiency, reduces heat gain, minimizes solar glare, eliminates UV and infrared light penetration, reduces HVAC costs, made with recyclable materials

TESTS / EXAMINATIONS
Listing available on request

LIMITATIONS
Reduced visibility

CONTACT
Duo-Gard Industries Inc.
40442 Koppernick Road
Canton, MI 48187
Tel: 734-207-9700
www.duo-gard.com
info@duo-gard.com

INTEGRAL LIGHT-MODULATING WINDOW SHADING

Hoberman Associates developed Adaptive Fritting in order to imbue an established architectural treatment with expanded functionality. Similar to standard fritted glass, this invention utilizes a graphic-pattern surface treatment in order to control heat gain and modulate light, while allowing sufficient transparency for viewing. However, while conventional fritting relies on a fixed pattern density, Adaptive Fritting provides a variable density that can modulate the amount of light transmittance. This performance is achieved by shifting a series of transparent fritted layers so that the graphic pattern can transition between a fully aligned state (minimal coverage) and an offset state (maximum coverage). In one installation, the panels are programmed to demonstrate the dynamic responsiveness of a field where light transmission and views can continuously adapt and change. As the frits transform, the visual effect is of sparse dots blossoming into an opaque surface.

CONTENTS
Cast acrylic, aluminum

APPLICATIONS
Control of solar gain and light levels, resulting in decreased energy consumption and improved occupant comfort

TYPES / SIZES
Design enables high degree of flexibility; built example consists of six 4' x 4' x 1" (1.2 m x 1.2 m x 2.5 cm) panels built into a curved wall

ENVIRONMENTAL
Provides modulation between interior and exterior space; creates variable degrees of shading controllable in real time

CONTACT
Hoberman Associates
40 Worth Street, Suite 1680
New York, NY 10013
Tel: 212-349-7919
www.hoberman.com
associates@hoberman.com

TRANSFORMATIONAL PRODUCT

Aire Pad

AIR-FILLED RUBBER PAD

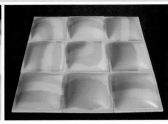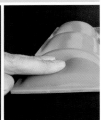

Developed by designer Fabrice Covelli, Aire Pad applies Fproduct's gel-encapsulation technology to noncompressed air trapped within a strong flexible skin. The enclosed air provides an efficient and comfortable cushioning effect; meanwhile, the flexible skin retains its sleek appearance after pressure is released. Products made with this technology are lighter, faster to produce, and less expensive than gel-filled items. Rubber and silicon are ideal materials for this application, and the color, mold pattern, and embossed design may be customized. Although the skin has good cut resistance, the material should not be installed in high-traffic public areas.

CONTENTS
Rubber or silicon, air

APPLICATIONS
Wall tiles and panels, cushions, furniture

TYPES / SIZES
Maximum size 2 x 3'
(.6 x .9 m)

ENVIRONMENTAL
Efficient use of material

TESTS / EXAMINATIONS
UV resistant, waterproof, good shock and sound absorption

LIMITATIONS
Cushioning effect will be nullified if punctured

CONTACT
Fproduct
250 Saint Marks Avenue
Brooklyn, NY 11238
Tel: 917-202-2349
www.fproduct.net
getinfo@fproduct.net

PHOTOLUMINESCENT AGGREGATES

 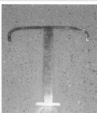 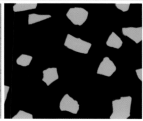 

Ambient Glow Technology (AGT), created by Peter Tomé, is a proprietary blend of nonshrinking, extremely hard polyester resin and photoluminescent pigment. Exhibiting very high luminosity, AGT was specifically formulated for use in decorative concrete, stucco, plasters, cementitious overlays, terrazzo, and epoxy applications. AGT aggregates produce an unusual ambient light source, enhancing safety in low light-level conditions for over twelve hours after exposure to sunlight for only ten minutes. The application of AGT in exterior landscaping projects can actually reduce the need for electrically powered lighting by up to 70 percent per evening. AGT's daylight or "native" color is off-white, and afterglow colors are available in yellow-green, aqua blue, and sky blue.

CONTENTS
Europium-doped, strontium aluminate photo-luminescent pigment combined with polyester resin

APPLICATIONS
Exterior: patio stones, pavers, stair treads, furniture, walkways, pool surrounds, decks; interior: floors, tiles, countertops, bathroom vanities, fireplace surrounds, walls, stair treads

TYPES / SIZES
Fine sand, small sand, large sand; 1/4" and 1/2" stones; available in yellow-green, aqua-blue and sky-blue afterglow

ENVIRONMENTAL
Reduced electricity use, reduced light pollution, renewable ambient light source

TESTS / EXAMINATIONS
Aggregates work extremely well between 482°F and -4°F (250°C and -20°C); AGT can endure strong UV radiation and does not discolor under 300-watt high mercury lamp for 1,000 hours at 95–104°F (35–40°C), 80% humidity; AGT is stable with use in combination with water, stains, acids, epoxies, etc.

LIMITATIONS
Performance limited by the level of adjacent competing light sources

CONTACT
Ambient Glow Technology
1898 Liverpool Road
Pickering, ON L1V 1W5
Canada
Tel: 905-250-9642
www.ambient
glowtechnology.com
ptome@ambient
glowtechnology.com

ULTRAPERFORMING MATERIAL

Bicicleta

RECYCLED BICYCLE INNER TUBE RUG

Barcelona-based Nanimarquina addresses the problem of waste directly in their carpets. Bicicleta gives a second life to discarded bicycle inner tubes from the city of Panipat, in northern India, where this rug is produced. Its ingenious hand-crafted design makes it possible to transfer rubber from the garbage of village streets to the living rooms, studios, and terraces of houses in various parts of the world. Created by designers Nani Marquina and Ariadna Miquel, Bicicleta requires between 130 and 140 bicycle inner tubes that, once collected, are washed, cut, and woven on a loom.

CONTENTS
100% recycled rubber

APPLICATIONS
Floor covering

TYPES / SIZES
5.6 x 7.9' (1.7 x 2.4 m); loop length 1.4" (35 mm); weight 2.05 lbs/ft^2 (10 kg/m^2)

ENVIRONMENTAL
100% recycled post-consumer waste, mitigates problematic trash heaps

LIMITATIONS
Hand wash only

CONTACT
Nanimarquina
Església 10, 3er D
Barcelona, 08024
Spain
Tel: +34 932-376-465
www.nanimarquina.com
info@nanimarquina.com

Bio-Based Foams

BIO-BASED FLEXIBLE AND RIGID FOAMS

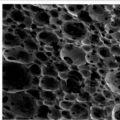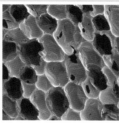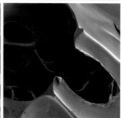

Richard P. Wool of the Center for Composite Materials has synthesized flexible/rigid polyurethane foams from soybean oil polyols. The advantage of these foams is that they can replace petroleum-based materials (synthetic polyols) and can be used for many applications. The morphology of the foams can be controlled by several factors: the type and functionality of the soybean oil polyols, the type of curing agents, the amount of water, and the amount of catalyst. Both flexible and rigid foams can be developed from vegetable oils by implementing different processes. The biocontent of the foams varies from 33 to 96 weight percent. The use of this biodegradable, locally harvested, and renewable source has economic and environmental advantages that make it an attractive alternative to petroleum-based materials.

CONTENTS
Chemically modified soybean oil

APPLICATIONS
Footwear, construction, cars, electronics, sporting goods, and many other uses

ENVIRONMENTAL
Use of renewable resources, biodegradable

TESTS / EXAMINATIONS
ASTM D1621, D1622, D3574, ASTM D3576

LIMITATIONS
Relatively expensive

CONTACT
University of Delaware
Center for Composite
Materials
Newark, DE 19716-3144
Tel: 302-831-3312
www.che.udel.edu/
research_groups/wool/
wool@udel.edu

ClearSeam ITL

SELF-STRUCTURAL TRANSLUCENT WALL SYSTEM

Panelite has recently developed the ClearSeam ITL fabrication system, a clear polycarbonate spine for interlocking self-structural translucent wall panels on-site. Whereas the original ITL system uses a brushed aluminum, tongue-and-groove detail, ClearSeam ITL delivers unprecedented lightness and transparency.

ClearSeam ITL features the same labor-saving and self-supporting attributes as the original tongue-and-groove option. Both are compatible with Panelite's Cast Polymer Series honeycomb panels, which can be specified flat or curved, clear or colored.

All Panelite ITL Systems are prefabricated according to the project's specifications and arrive on-site ready to install. Shop drawings are included and Panelite technical support is available from design development through installation.

CONTENTS
Panelite cast-polymer panels with integrated aluminum profiles

APPLICATIONS
Partitions, feature walls, backlit wall systems, exhibitions

TYPES / SIZES
1" (2.5 cm) and 1.5" (3.8 cm) thick wall systems available

ENVIRONMENTAL
Minimizes or eliminates the necessity for additional structural framing members, effectively reduces waste upon demolition

TESTS / EXAMINATIONS
Class C and B fire rated systems available per ASTM-E84

LIMITATIONS
Flat and curved systems available

CONTACT
Panelite
315 West 39th Street, Suite 807
New York, NY 10018
Tel: 212-947-8292
www.panelite.us
info@panelite.us

ENVIRONMENTALLY TUNED WALL SYSTEM

Cloak Wall, designed by Blair Satterfield and Marc Swackhamer, interrogates the residential wall. It is a full-scale wall prototype that explores energy conservation through alternative approaches to cooling, heating, ventilating, and lighting an inexpensive house. The wall is designed for quick assembly using stacked, high-strength, low-weight "exo-skeletal" bricks. A soft "intelligent quilt" is suspended from this rigid outer shell to form the wall's weather seal.

Cloak Wall is self-supporting, clamped to its foundation by vertical tension cables. The geometry of Cloak Wall's blocks allows them to slide along one another horizontally in order to adjust window size. Depending on orientation, one wall might have large, dilated openings to allow for more natural light or for views of a landscape. Another side might have constricted openings to block intense southern sun or restrict a view. Additionally, as Cloak Wall's panels slide along one another, the wall can grow or shrink in height to create a sloped roof surface for drainage. By incorporating color-shifting paint borrowed from the automotive industry, the wall surface is able to either reflect or absorb radiant heat depending on whether the sun is high or low in the sky.

The "intelligent quilt" of each wall forms an interactive weather seal to control temperature, humidity, weather, light levels, and desired views. Layers of the quilt keep water out, provide insulation, carry utility lines, and acoustically soften the interior. For example, luminescent fabrics woven into the quilt supply interior light, recharging passively during the day. Air-filled pockets of transparent ETFE polymer allow insulation levels to be fine-tuned depending on outside air temperature.

CONTENTS

Polymethyl methacrylate (PMMA), felt, ethylene tetrafluoroethylene (ETFE), multihued metallic automotive paint

APPLICATIONS

Interior and exterior residential wall system

TYPES / SIZES

Module size 34 x 16 x 12" (86.4 x 40.6 x 30.5 cm); 30 different colors ranging from greens and yellows at the base, to browns and grays along the horizon line, to blues and purples at the top

ENVIRONMENTAL

Interactive weather seal, reduction in energy consumption

LIMITATIONS

Only one story high

CONTACT

HouMinn Practice
6344 Warren Avenue South
Edina, MN 55439
Tel: 612-669-2603
www.houminn.com
marc@houminn.com

Dura

GRAPHICALLY ENHANCED ARCHITECTURAL ACRYLIC

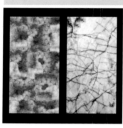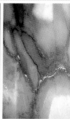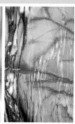

R-Cast Dura is a new Class A and Class I fire-rated material developed by Reynolds Polymer Technology, Inc. for architectural applications. The acrylic material is formable and bondable, allowing a wide variety of custom shapes. It also has a 5–10 percent light transmission, so backlighting produces a soft glow.

Architects and designers can select from a range of standard patterns and colors, or they may submit customized images or graphics to be reproduced on the material. Dura is well-suited for a variety of projects, such as display cases, point-of-purchase displays, and furniture.

CONTENTS
Methacrylate resin

APPLICATIONS
Paneling, artwork, displays; retail, hospitality, and commercial applications

TYPES / SIZES
Black opaque, white opaque, custom color/ image/graphic; thicknesses 1/4" (.6 cm) and 1/2" (1.3 cm); sheet size 4 x 8' (1.2 x 2.4 m); custom sheet sizes and thicknesses available

ENVIRONMENTAL
Manufactured in a facility that recycles waste and water

TESTS / EXAMINATIONS
ANSI Z124; ASTM-D570, D638, D696, D785, D790, D792-08, E84; G21, G22; NEMA LD-3

CONTACT
Reynolds Polymer Technology, Inc.
607 Hollingsworth Street
Grand Junction, CO 81505
Tel: 800-433-9293
www.reynoldspolymer.com
customerservice@ reynoldspolymer.com

Expancel

THERMOPLASTIC MICROSPHERES

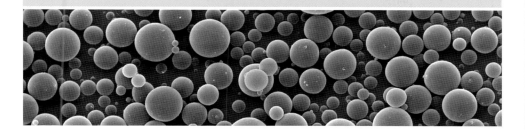

Unexpanded Expancel thermoplastic microspheres are small plastic particles that contain a droplet of liquid hydrocarbon. When heated, the plastic shells soften and the liquid builds pressure, causing the particles to expand. The result is gas-filled, thin-walled particles of extremely low density. The particles are compressible and resilient, and these properties are transferred to the material or component to which they have been added. The closed-cell microspheres are good thermal and acoustic insulators.

Expancel is available in expanded or unexpanded forms for different applications. In its heat-activatable, unexpanded form Expancel is used as a foaming agent. The expanded form of Expancel is an ultralow-density filler.

CONTENTS
Thermoplastic resin shell, hydrocarbon-filled core

APPLICATIONS
Expandable coatings and gap-filling adhesives, intumescent coatings and sealants, puff ink, extruded and molded plastic items (unexpanded); resilient filler for thermoset polyester resins, compressible elastomers, vibration-damping materials, paint, roof coatings, adhesives, sealants, prosthetics (pre-expanded)

TYPES / SIZES
DE (dry expanded), WE (wet expanded), DU (dry unexpanded), WU (wet unexpanded), MB (palletized masterbatch of DU); particle diameters from 12 to 120 microns

ENVIRONMENTAL
Weight reduction reduces material use and transport costs; insulating properties

LIMITATIONS
Thermoplastic composition limits maximum temperature exposure

CONTACT
Expancel/
Eka Chemicals Inc.
2240 Northmont Parkway
Duluth, GA 30096
Tel: 678-775-5102
www.expancel.com
info.expancel@
akzonobel.com

Flowerfall

TRANSLUCENT SCREEN MADE FROM POSTCONSUMER WASTE

Michelle Brand designed Flowerfall specifically to address our ever-growing amount of plastic thrown away each year. Flowerfall is an eco-contemporary "curtain" made of the bases of discarded PET bottles and articulated tagger ties, the plastic ties typically used to connect a price tag to an item. In her construction of the soft screens, Brand pays great attention to minimizing processing, and uses no melting, reforming, or molds in her handmade fabrication process. Flowerfall celebrates both form and function in sculptural installations that are decorative without being overly fussy. The material has an unexpected softening quality that can counter hard surfaces or spaces, and its translucent properties harness and filter light in dramatic ways.

CONTENTS
100% polyethylene
terephthalate (PET)

APPLICATIONS
Chandeliers, window
screens, room dividers
(interior); garden rooms,
decorative canopies,
sculptural installations
(exterior)

TYPES / SIZES
Cascade, Mini Cascade,
Flowerfall, Cascade
Lancashire

ENVIRONMENTAL
Second-use material, locally
sourced, closed-loop
collection and process
system, 100% recyclable

CONTACT
The Greenhaus Ltd.
Tel: +44 (0)1279-658400
www.michellebrand.co.uk
jane@thegreenhaus.co.uk

ADAPTABLE SILICONE MATERIAL

 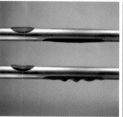 

Formerol is a new class of silicone materials that combines the adhesive and room-temperature-curing properties of silicone adhesives with the moldability of silicone rubber. Benefits include excellent adhesion to aluminum and other metals, which allows soft-touch silicone inserts or surface coatings to be molded directly onto these substrates without adhesives. Additionally, Formerol's ability to be molded and cured at room temperature—with excellent adhesion to unusual materials such as wood, ceramics, and glass—allows soft-touch textured surfaces to be applied in new and unconventional ways.

Because Formerol can be cured at room-temperature, inserts and parts may be designed that conform to individual end users' personal requirements. This flexible technology allows for new opportunities in applications ranging from sports equipment to healthcare.

CONTENTS
Silicone, fillers, functional additives

APPLICATIONS
Soft-touch molding at ambient temperatures, grip and insert molding directly onto aluminum parts, mass customization of grips

TYPES / SIZES
Custom

ENVIRONMENTAL
Durable in extreme weather conditions, excellent UV resistance, electrically insulating

TESTS / EXAMINATIONS
ASTM D 926, ASTM D638, BS 903 part A21

LIMITATIONS
Recommended molding temperature 32–70°F (0–21°C), recommended curing temperature 68–140°F (20–60°C); atmospheric humidity required for cure

CONTACT
FormFormForm Ltd.
13 Hague Street
London, E2 6HN
United Kingdom
Tel: +44 (0)20-7739-9446
www.formerol.com
info@formerol.com

Rubber

...TIVE SURFACE PANELS WITH APPLIED GRAPHICS

INTERFACIAL PROCESS

Decorative Graphic Panels are created by a process that incorporates a translucent and/or graphic material applied to a variety of substrates, suitable for commercial interiors. Potential applications include partitions, doors, countertops (bar, vanity, and furniture tops), point-of-purchase displays, visual merchandising, and light diffusers. Graphic Panels are offered in seven hundred standard patterns with custom imaging available. Surfaces may include glass and a range of plastics, and may be specified as low-iron, textured, frosted, overlaminated (matte or glossy), or double laminated, with custom edgework possible.

CONTENTS
Glass, acrylic, polycarbonate, polyethylene terephthalate glycol (PETG) or cellulose acetate (CAB)

APPLICATIONS
Vertical and horizontal surfaces in commercial, retail, and hospitality functions

TYPES / SIZES
4 x 8' (1.2 x 2.4 m), CAB 4 x 6.3' (1.2 x 1.9 m); thickness .12" (3 mm) or .24" (5.9 mm), glass .125" (3.2 mm) or .25" (6 mm) only, CAB .1" (2.5 mm) only

ENVIRONMENTAL
PETG panels made from recycled content; CAB made from renewable wood pulp

LIMITATIONS
Interior use only

CONTACT
Architectural Systems, Inc.
150 West 25th Street,
8th Floor
New York, NY 10001
Tel: 800-793-0224
www.archsystems.com
sales@archsystems.com

IVY-INSPIRED SOLAR AND WIND ENERGY CURTAIN

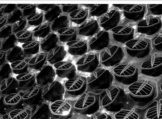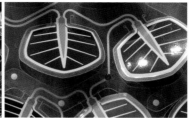

Grow is a hybrid energy-delivery device inspired by ivy. Grow's "leaves" are flexible, organic photovoltaic panels that capture solar energy and convert it into electricity. Each leaf is attached by a robust piezoelectric generator at the leaf's stem. When the leaves flutter in the wind, the stems flex to produce electricity while also creating a provocative kinetic experience.

Grow's modular system is designed to be attached to any surface of a building that receives sunlight and wind. Grow designers SMIT utilize recycled and reclaimed materials wherever possible and are committed to sustainable methods of recycling in order to minimize the product's environmental footprint at the end of its lifespan.

TRANSFORMATIONAL PRODUCT

CONTENTS
Konarka Power Plastic encapsulated in ethylene tetrafluoroethylene (ETFE) fluoropolymer lamination, macrofiber composite-piezoelectric generator (with printed circuit boards and energy-harvesting electronics), aluminum standoffs, stainless-steel fastening hardware, copper core wire, and rubber gaskets

APPLICATIONS
Solar and wind power generation

TYPES / SIZES
16 x 12 x 3" (40.6 x 30.5 x 7.6 cm) single module; multiple modules connect together to become larger systems

ENVIRONMENTAL
Renewable energy production, recyclable/ reclaimable materials, efficient use of materials

LIMITATIONS
Limited life span (15 to 20 years)

CONTACT
SMIT
63 Flushing Avenue, Unit 195, Building 280, Suite 515
Brooklyn, NY 11205
Tel: 718-399-4452
www.s-m-i-t.com
contact@s-m-i-t.com

Heat Treated Carpet

REUSED CARPET SHEETING MATERIAL

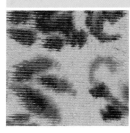  

Carpet is typically used for seven to ten years (residential) or five to eight years (commercial), and recarpeting accounts for 55 percent of all carpet sold. Moreover, carpet has been estimated to require up to two hundred and fifty thousand years to biodegrade. Heat Treated Carpet, designed by Kelly Atkins, is a sheet material produced from polypropylene-based carpet waste. Manufactured by Carpet-Burns Ltd., the product is a plywood-style rigid board made from carpet cutoffs, and the colors and patterns of the original carpet remain visible through the surface of the material. Heat Treated Carpet may be drilled, sawn, or molded, and is waterproof, durable, nonporous, and very resistant to staining. In addition, the material is suitable for interior and exterior applications.

CONTENTS
Polypropylene-based carpet

APPLICATIONS
Kitchens, bathrooms, bars, restaurants, flat-packed furniture, flooring (screed and tile), staircases, shop fittings, seating, architectural features, paneling, wall tiles, molded consumables

TYPES / SIZES
Available in various thicknesses, including .12", .35", .47", .59", and .71" (3 mm, 9 mm, 12 mm, 15 mm, and 18 mm)

ENVIRONMENTAL
100% recycled; no stabilizers, glues, or resins are added

CONTACT
Carpet-Burns Ltd.
Banksmill Studios,
71 Bridge Street
Derby, DE1 3LB
United Kingdom
Tel: 01332-297734
www.carpet-burns.com
kelly@carpet-burns.com

Ice

ICE-MIMICKING ACRYLIC

Formed out of Reynolds Polymer Technology's R-Cast Acrylic product line, the texture of Ice is completely randomized without any repetitive patterns. This randomization ensures that the acrylic Ice more closely resembles real ice, particularly with light refractions off the surface.

R-Cast Ice is ideal for use within the hospitality and entertainment markets. Each sheet of Ice is unique, formable, bondable, available to cut to any size or shape, and can have a striking impact with special lighting effects, making it perfect for projects like restaurants, hotels, or nightclubs.

By employing a variety of different translucencies and hundreds of available colors—as well as the creative use of lighting—innumerable effects can be realized with R-Cast Ice. Available in half-inch and one-inch thicknesses, R-Cast Ice is available in sheets as large as 4 x 8 feet (1.2 x 2.4 meters), which can be fabricated or formed into a variety of shapes for custom projects.

CONTENTS
Polymethyl Methacrylate
(PMMA)

APPLICATIONS
Decorative paneling,
partitions, water features,
signage, furniture, displays

TYPES / SIZES
Sheet size 4 x 8' (1.2 x 2.4
m); thickness 1/2" (1.3 cm)
and 1" (2.5 cm); clear,
frosted, or custom color

ENVIRONMENTAL
Manufactured in a facility
that recycles waste and
water

TESTS / EXAMINATIONS
ASTM-D256, D570, D621,
D638, D648, D695, D696,
D732, D785, D790, D792

LIMITATIONS
Not available with material
less than 1/2" (1.3 cm) thick

CONTACT
Reynolds Polymer
Technology, Inc.
607 Hollingsworth Street
Grand Junction, CO 81505
Tel: 800-433-9293
www.reynoldspolymer.com
customerservice@
reynoldspolymer.com

Jali Cascata

LIGHT-REFRACTIVE POLYMER SANDWICH PANEL

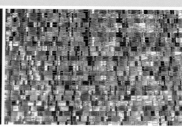

Sensitile Systems's Jali Cascata panels, designed by Abhinand Lath, combine the light-refractive properties of their Jali series of panels—acrylic-based composites designed to harness and extend the intensity of ambient illumination—with the color dispersion of a dichoric film layer. The result is a visually arresting array of multiple colors that dynamically respond to changing viewing angles, producing unexpected changes in tandem with contextual lighting conditions.

CONTENTS
polymethyl methacrylate (PMMA), dichroic film

APPLICATIONS
Vertical and horizontal surfacing

TYPES / SIZES
48 x 96 x 3/4" (122 x 244 x 1.9 cm)

TESTS / EXAMINATIONS
ASTM D785, D696, 1929; others available upon request

LIMITATIONS
Presently cannot be made larger than 48 x 96" (122 x 244 cm) without a seam

CONTACT
Sensitile Systems
1735 Holmes Road
Ypsilanti, MI 48198
Tel: 313-872-6314
www.sensitile.com
info@sensitile.com

Kaynemaile

POLYMERIC SEAMLESS MESH

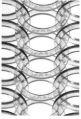

Kaynemaile is a seamless mesh composed of polymer rings. The advantage of the Kaynemaile process is its ability to create multiple elements and assemble them into a three-dimensional structured mesh at the same time, as opposed to molding components first and assembling them later. Every single component interconnects with at least four other parts, without the need for additional linking materials or adhesives.

Kaynemaile's seamless mesh can actually be made from many different materials—including liquid-crystal polymers, polycarbonates, thermoset urethanes, rubber, and metal alloys—the requirement being whether a material can phase change from a liquid to a solid state. As a modern alternative to traditional woven, welded, or perforated metal sheeting, Kaynemaile is lighter, price-competitive, has faster turnaround and delivery, efficient to freight (one-eighth the embodied energy by weight), easier to install, changeable in appearance through the use of lighting, and constantly evolving to accommodate future trends through the use of low-environmental-impact polymers. Kaynemaile's oil spill recovery attributes use the three-dimensional structure of the mesh to capture and contain heavy crude oil when floating on saltwater or freshwater environments, which is then removed through a simple process and redeployed for continuous recovery.

CONTENTS
Engineering-grade, UV-stabilized polycarbonate

APPLICATIONS
Feature lighting, safety barriers, suspended ceiling installations, security screening, solar screening, spatial division, and curtains

TYPES / SIZES
ø .87" (2.2 cm) ring size

ENVIRONMENTAL
100% recyclable, manufactured via zero-waste injection-molding technique, material recovery and reprocessing offered, efficient use of material

TESTS / EXAMINATIONS
AS1530 Flammability Index of (FI)=6; IMO resolution A.652(16), 1989 fire test procedure; UL94 V-O at 1.5mm self-extinguishing; ISO 15025 2000 (E)

LIMITATIONS
In exterior applications, color stability is limited to 8 years and structural integrity maintains 80% of tensile strength after 10 years

CONTACT
Kaynemaile Ltd.
Unit Two 85 Hutt Road
Thorndon, Wellington 6135
New Zealand
Tel: +64 4-473-4989
www.kaynemaile.com
info@kaynemaile.com

Kozo

DECORATIVE TRANSLUCENT ACRYLIC PANELS

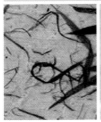

Kozo decorative surface panels feature vibrant natural fibers encapsulated in translucent acrylic, and are ideal for backlit vertical installations. Potential applications include store fixtures, feature walls, cabinetry, trim, signage, and thermoformed shapes. The Kozo collection is offered in sixteen dynamic colors and patterns with a gloss standard finish (matte available).

CONTENTS
Acrylic panels, encapsulated natural fibers

APPLICATIONS
Retail, commercial, and hospitality applications; ideal for backlit vertical applications

TYPES / SIZES
Panel size 4 x 8' (1.2 x 2.4 m); thicknesses: 1/16", 1/8", 3/16", 1/4", 3/8", 1/2", 3/4", and 1" (1.5 mm, 3.2 mm, 4.8 mm, 6 mm, 9.5 mm, 13 mm, 19 mm, and 25 mm)

LIMITATIONS
Interior use only

CONTACT
Architectural Systems, Inc.
150 West 25th Street,
8th Floor
New York, NY 10001
Tel: 800-793-0224
www.archsystems.com
sales@archsystems.com

Lunalite

DECORATIVE INTERLAYER SURFACE PANELS

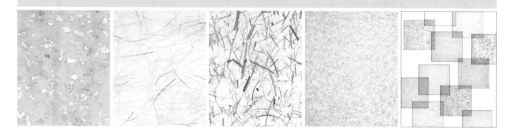

Lunalite decorative surfaces exhibit a range of effects, from rustic textures to metallic shimmer. This collection of sophisticated interlayers features materials such as mica, gold flake, and sisal fibers, giving surfaces depth and complexity. Custom fabrication ensures that each organic Lunalite surface is unique. Natural, metallic, holographic, and recycled inclusions are embedded in a range of materials, including glass and acrylic, and glass panels may be tinted to match virtually any color. Lunalite technology can be applied to doors, windows, dividers, walls, countertops, displays, and more.

CONTENTS
Glass, polycarbonate, acrylic, polyethylene terephthalate glycol (PETG) panels; mica, gold flake and sisal fibers as interlayers

APPLICATIONS
Decorative surfacing applications in all commercial environments, including doors, windows, dividers, walls, countertops, and displays

TYPES / SIZES
Size 4 x 8' (1.2 x 2.4 m); thickness: acrylic 1/8" (3.2 mm), 1/4" (6 mm); glass 1/2" (13 mm); polycarbonate 1/16" (1.5 mm), 1/8" (3.2 mm), 1/4" (6 mm); PETG 1/8" (3.2 mm), 1/4" (6 mm)

ENVIRONMENTAL
PETG panels and some inclusions are made from recycled content

LIMITATIONS
Interior use only

CONTACT
Architectural Systems, Inc.
150 West 25th Street,
8th Floor
New York, NY 10001
Tel: 800-793-0224
www.archsystems.com
sales@archsystems.com

Memorial Rebirth

ATMOSPHERIC THICKENING

Developed by Japanese artist Shinji Ohmaki, Memorial Rebirth is an installation intended to "weave stories with countless bubbles." The project consists of fifty machines capable of producing fifty thousand bubbles per minute, placed in several strategic public locations. Demonstrating the use of minimal material for maximum perceptual effect, Memorial Rebirth produces a thick atmosphere of bubbles that suddenly appear in an urban space and disappear rapidly. This spectacle is designed to represent the "frequency of vital energy," and to provide viewers with the sense of being drawn into an "energized" space. Ohmaki states that once viewers experience Memorial Rebirth, "It will appear to the viewers as if everything they see is ephemeral as bubbles."

CONTENTS
Bubble liquid, fiber-reinforced plastic (FRP), aluminum, bubble machine

APPLICATIONS
Public art

TYPES / SIZES
Each machine ø 19.9" (50.4 cm), 15.8" (40 cm) high

LIMITATIONS
Not for use in bad weather

CONTACT
Tokyo Gallery + BTAP
7F, 8-10-5 Ginza, Chuo-ku
Tokyo, 104-0061
Japan
Tel: +81 3-3571-1808
www.tokyo-gallery.com
info@tokyo-gallery.com

STATIC ELECTRICITY–TRACING SURFACE

B.lab Italia's Metal Series surfaces contain liquids infused with metal particles that are trapped within a resilient polymer shell. The result is an interactive floor tile or horizontal surface that responds to pressure as well as electric charge. Building occupants' footsteps are temporarily recorded by the movement of fluids, and a high-level static charge will actually allow a prolonged "drawing" to be made within the surface. According to designers Gianfranco Barban and Gregg Brodarick, B-Surfaces Metal Series are intended to transform movement into memory, appropriating the floor as a tool for creative expression. The liquids contained within Metal Series are nontoxic, and tiles should be laid on smooth horizontal surfaces for best effect.

CONTENTS
Polycarbonate (PC), polyvinyl chloride (PVC), polyethylene terephthalate (PET), nontoxic liquids

APPLICATIONS
Flooring for residential and commercial interiors

TYPES / SIZES
Tiles 39 x 39" (99.1 x 99.1 cm) and 19 1/2 x 19 1/2" (49.5 x 49.5 cm)

ENVIRONMENTAL
Nontoxic liquids

TESTS / EXAMINATIONS
Certified flooring product for residential and commercial interiors, Class 1 fire rating

LIMITATIONS
Interior use only

CONTACT
B.lab Italia
Via Marmolada 20
Gallarate, VA 21052
Italy
Tel: +39 0331-774445
www.blabitalia.com
bmail@blabitalia.com

TRANSFORMATIONAL MATERIAL

Mirage

GRAPHICALLY ENHANCED ACRYLIC

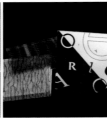

R-Cast Mirage is a custom acrylic panel for interiors. Custom colors, images, or graphics can be accurately reproduced using specially developed fabrication methods, in order to create particular visual effects within interior environments. Reynolds Polymer Technology offers a number of library images categorized by natural, industrial, and abstract themes, as well as the possibility for image customization. At half the weight and seventeen times the strength of glass, Mirage is formable, bondable, and can be fabricated into various shapes.

CONTENTS
Polymethyl methacrylate (PMMA)

APPLICATIONS
Decorative paneling, partitions, receptions, door and cabinet inserts, artwork, signage, lighting, and furniture

TYPES / SIZES
Sheet sizes 4 x 8' (1.2 x 2.4 m) and 4 x 10' (1.2 x 3 m); thickness 1/8", 1/4", 3/8", 1/2", 3/4", and 1" (3.2, 6, 9.5, 13, 19, and 25 mm); white opaque, white translucent, and clear acrylic

ENVIRONMENTAL
Manufactured in a facility that recycles waste and water

TESTS / EXAMINATIONS
ASTM-D256, D570, D621, D638, D648, D695, D696, D732, D785, D790, D792

LIMITATIONS
Indoor applications only

CONTACT
Reynolds Polymer Technology, Inc.
607 Hollingsworth Street
Grand Junction, CO 81505
Tel: 800-433-9293
www.reynoldspolymer.com
customerservice@
reynoldspolymer.com

Mycoply

MYCOLOGICAL BIOMATERIAL

Mycoply board, designed by Edward Browka, is a mycological biomaterial that is actually grown in a few months. It has properties similar to balsa wood and can be easily formed when saturated, yet will hold its shape once dried. Mycoply is rapidly renewable, grown using agricultural byproducts, and does not interrupt its surrounding ecosystem. Mycoply can replace core materials such as balsa, honeycomb, and a variety of petroleum-based foams, and can be used for wind turbine blades, boat hulls, lightweight vehicle panels, and many other applications. Mycoply also uses a low-embodied-energy manufacturing process, as the material self-assembles at room temperature and pressure in the dark. Additionally, its performance is superior in many ways to current structural core materials due to the ease in which it may be bent and formed. Mycoply reduces waste as it is grown to near-net shape and does not require postprocessing.

CONTENTS
Chitinous polymer

APPLICATIONS
Boat hulls, wind turbine blade cores, lightweight vehicle panels

TYPES / SIZES
Size varies according to application; can be grown to near-net shapes; can be formed at 30% moisture; holds form when dried

ENVIRONMENTAL
100% natural, rapidly renewable (grows in months), biodegradable, low-embodied-energy manufacturing process, grown from waste materials, locally produced

TESTS / EXAMINATIONS
ASTM Standards: D3501-05a: Standard Test Methods for Wood-Based Structural Panels in Compression; D3500-90: Standard Test Methods for Structural Panels in Tension; D3043-00: Standard Test Methods for Structural Panels in Flexure

CONTACT
Ecovative Design
1223 Peoples Avenue
Troy, NY 12180
Tel: 518-690-0399
www.ecovativedesign.com
info@ecovativedesign.com

MYPODLife

ILLUMINATED LIVING WALL

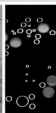

MYPODLife, designed by Freya Bardell and Brian Howe, is a living wall and light box made of DuPont Corian. Part of an ongoing investigation into living wall technologies, MYPODLife operates as an air filter, improving indoor environmental quality and removing harmful volatile organic compounds (VOCs). In this way, the system serves as a functional alternative to traditional wall art. The surface of MYPODLife is etched on the front and back, revealing design layers when illuminated. Plant pods are used to germinate highly oxygenating indoor plants, including shade-loving herbs, and are watered using waste tea water.

CONTENTS
Soil, plants, Corian, LED
strip lights

APPLICATIONS
Green wall, garden, air
purifier

TYPES / SIZES
3 x 8' (.9 x 2.4 m); modular
dimensions variable;
available in any Corian
color

ENVIRONMENTAL
Improved indoor air quality

LIMITATIONS
Each piece is custom
designed

CONTACT
Greenmeme
1133 Isabel Street
Los Angeles, CA 90065
www.greenmeme.com
contact.greenmeme@gmail
.com

Narco

ROTOMOLDED SLEEP CHAMBER

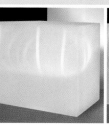

Narco is a cell for napping. Developed by Marie-Virginie Berbet for use in the workplace, Narco optimizes environmental conditions to provide a regenerating short-time sleep by preventing sleep inertia (which is characterized by lower motor dexterity and a feeling of grogginess).

The monolithic rotomolded shell allows an unobtrusive integration of the cell into the workplace and direct access to sleep. Its translucent exterior provides audio and visual isolation from its outer environment without conveying a claustrophobic feeling. This isolation is reinforced inside by the densification of foam strips around the head of the user. The hammock shape of the interior cocoon allows a levitation position, which reduces sleep latency but also sleep depth. Thanks to breathing rhythm sensors placed on the user's body, Narco detects precisely when the napper falls asleep. Ten minutes later—the optimal time for avoiding sleep inertia—a soft light increases in the cocoon strips in order to awake the user gently.

CONTENTS
Rotomolded shell, foam strips, metal skeleton, LEDs, pressure sensor

APPLICATIONS
Napping

TYPES / SIZES
5.9 x 4.3 x 5.2' (1.8 x 1.3 x 1.6 m)

ENVIRONMENTAL
Enhanced occupant health

CONTACT
Marie-Virginie Berbet
5, Square Trudaine
Paris, 75009
France
Tel: +33 610-191-573
www.mvberbet.com
mvberbet@mvberbet.com

Pneu-Green Facade

GARDEN FACADE WITH PNEUMATIC OUTER SKIN

Experimonde has developed a new kind of membrane curtain wall for domestic retrofits. Integrating the idea for an insulating skin with a garden, the Vienna University of Technology–based team recognized that the buffer zone between the two layers in a double-cushion pneumatic system could be used as a green house.

The Pneu-Green Facade is an inflatable cladding system designed to mitigate thermal transfer while providing a large three-dimensional matrix for a domestic garden. Garden plots are located near windows for easy access, and horizontally sliding shelf systems allow the garden to extend greater distances from openings.

CONTENTS
Ethylene tetrafluoroethylene (ETFE) membrane cushion, steel frame construction, sliding planter shelves

APPLICATIONS
Climate control, noise reduction, dust reduction, wind protection, lighting enhancement

TYPES / SIZES
Minimum dimensions 47.3 x 47.3" (120 x 120 cm); custom sizes available

ENVIRONMENTAL
Reduced operating energy, recyclable materials, active oxygenation

LIMITATIONS
Prototype phase

CONTACT
Experimonde
Arsenal 210,
Franz-Grill-Strasse 3
Vienna, 1030
Austria
Tel: +43 (0)6888-19-19-66
www.experimonde.eu
pmschultes@
experimonde.eu

Pneumocell

INFLATABLE CELL BUILDING SYSTEM

Pneumocell is an assembly kit consisting of inflatable building elements analogous to biological cell-structures, which can be connected in numerous combinations to form complete constructions. By specifying a common polygonal edge-length for all of the cells, designer Thomas Herzig has ensured that all element types fit together precisely. The cells are airtight and do not require constant inflation, as in other air-supported structures. If one element is damaged, the other elements can still give support to the construction, and the damaged element can be replaced—much like cells in a biological organism.

CONTENTS
Polyurethane or Polyvinyl
Chloride (PVC)

APPLICATIONS
Mobile lightweight
buildings

TYPES / SIZES
5.1–14.1' (1.55–4.30 m) cell
diameter; custom elements
possible

ENVIRONMENTAL
Minimal use of material,
100% recyclable

TESTS / EXAMINATIONS
Flame retardant B1 DIN
4102 / Önorm3800

LIMITATIONS
UV resistance not yet
certified

CONTACT
Thomas Herzig
Schluesselgasse 8
Vienna, 1040
Austria
Tel: +43 699-11-10-12-20
www.pneumocell.com
pneumocell@gmx.at

Rainwater H₂OG

LLDPE RAINWATER STORAGE MODULE

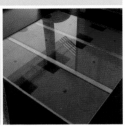

Essentially a water-filled building block, Rainwater H₂OG is a fifty-one-gallon reuseable, food-grade water storage module that works horizontally or vertically to provide rainwater or gray water storage either within a building structure or in tight spaces in and around a structure. The H₂OG module is sized to fit between joists and studs spaced at twenty-four-inch (sixty-centimeter) centers, enabling the stored water to be used as effective thermal mass within the building envelope. The H₂OG thus multitasks to provide a facility for water reuse as well as thermal mass for energy savings. Rainwater H₂OG is made of UV-stabilized LLDPE with brass-threaded fittings. U.S.-made H₂OGs are available made of fifteen percent recycled material, while in Australia all H₂OGs are legally required to be made of virgin material. H₂OGs are designed to ship naked, and to easily recycle at the end of their twenty-year predicted life.

CONTENTS

Linear low-density polyethylene (LLDPE): 100% for potable applications, or 15% recycled content for irrigation and gray-water use

APPLICATIONS

Rainwater and gray water storage, insulating applications

TYPES / SIZES

Single module 20 x 9.5 x 71" (50.8 x 24.1 x 180.3 cm); weight 40 lbs (18.1 kg)

ENVIRONMENTAL

Rainwater and gray water storage, thermal mass, recyclable

TESTS / EXAMINATIONS

Designed to Australian standards and tested with FSA by CSIRO in Australia; FDA-approved materials

LIMITATIONS

Lifespan shortened in direct sunlight; recycled content version not FDA approved for potable use

CONTACT

HOG Works Pty. Ltd. 402 Redwood Avenue Corte Madera, CA 94925 Tel: 415-891-8748 www.rainwaterhog.com info@rainwaterhog.com

Scintilla Lumina

LIGHT-EMITTING POLYMER PANEL

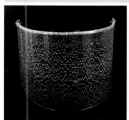 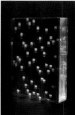 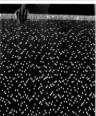  

Manufactured using Sensitile Systems' patented processes, Scintilla Lumina panels exploit the property of total internal reflection (TIR) to efficiently transport light from an edge condition (or other chosen location), emitting it from the panel face in a scattering of light points. The transfer of light within the material is so efficient that an 8-foot (2.4-meter) long panel can be illuminated from a single edge using a high-intensity LED source.

These panels, designed by Abhinand Lath, can be configured in a number of different ways in order to suit the application. For example, the light emission can be from one or both surfaces, the light terminals can be arranged into logos and graphics, and the panels can even be flexible in one dimension—allowing curvature down to an 18-inch (45.7-centimeter) radius.

CONTENTS
Polymethyl methacrylate
(PMMA), LEDs; glass
cladding also available

APPLICATIONS
Vertical and horizontal
surfacing

TYPES / SIZES
Custom-made to size up to
4 x 12' (1.2 x 3.7 m)

ENVIRONMENTAL
Efficient light diffuser

TESTS / EXAMINATIONS
ASTM D785, D696, 1929;
others available upon
request

LIMITATIONS
Presently cannot be made
thinner than 5/8" (1.6 cm)

CONTACT
Sensitile Systems
1735 Holmes Road
Ypsilanti, MI 48198
Tel: 313-872 6314
www.sensitile.com
info@sensitile.com

Self-Repairing and Sensing Matrices

MATERIALS THAT SELF-ADJUST TO INTERNAL STRESSES

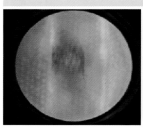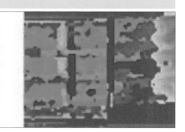

Since structural damage can go undetected in many materials, some products are often overengineered. However, materials that can provide information about their internal stresses—as well as trigger reliable self-healing properties—would allow manufacturers to be more confident in using lighter-weight materials. Self-Repairing and Sensing Matrices, designed by Carolyn Dry, can also survive longer than their conventional counterparts, thus reducing the use of virgin materials or petroleum-based resources. Natural Process Design has so far made airplane wings and other components of this material, which has been successfully tested for flexure, compression, and shear.

CONTENTS
Various

APPLICATIONS
Airplanes, ships, vehicles, infrastructure, buildings, sporting goods

ENVIRONMENTAL
Saves weight and fuel, offers assurance in composite use that self-repair has occurred

TESTS / EXAMINATIONS
Complete set of ASTM tests of flexure, compression, compression after impact, shear on self-repair

CONTACT
Natural Process Design Inc.
1250 East 8th Street
Winona, MN 55987
www.naturalprocessdesign
.com
naturalprocessdesign@
yahoo.com

Silva Cell

INTEGRATED TREE- AND STORMWATER-MANAGEMENT SYSTEM

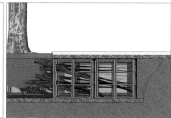

The Silva Cell integrated tree and stormwater system provides uncompacted soil volumes for large tree growth and on-site stormwater management while supporting traffic loads. The modular Silva Cells stack to create customized void spaces beneath paving such as sidewalks and plazas that are then filled with soil. This soil provides unlimited rooting space for trees, enabling them to grow large and improving canopy cover without disrupting surrounding utilities. The same soil volume simultaneously acts as on-site stormwater management, significantly reducing nonpoint source pollution and controlling the rate, volume, and quality of water runoff. By integrating trees, soil, and stormwater, the Silva Cell improves sustainable site design and brings enhanced ecological functionality to the built environment. The system meets engineering and loading standards with no loss of structural hard landscape integrity.

CONTENTS
Glass-reinforced polypropylene with galvanized steel tubes (product is 92% void space, to be filled with soil)

APPLICATIONS
Large urban tree growth and on-site stormwater management for streetscapes, plazas, parking lots, green roofs, and green walls

TYPES / SIZES
47.3 x 23.6 x 15.8" (120 x 60 x 40 cm); modular units can be stacked up to 3 high and spread laterally as wide as necessary to fit any site

ENVIRONMENTAL
Recyclable, enables large tree growth and effective on-site stormwater management in heavily paved areas

LIMITATIONS
Loading should not exceed AASHTO H-20 without consultation

CONTACT
DeepRoot
530 Washington Street
San Francisco, CA 94111
Tel: 415-781-9700
www.deeproot.com
info@deeproot.com

SITumbra

STRUCTURALLY INTEGRATED TRANSPARENT SOLAR-FACADE SYSTEM

 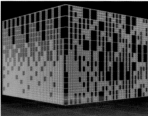 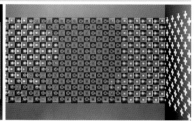

SITumbra is a passive solar-facade system that mediates between the seasons to reduce solar heat gain in summer and allow useful solar transmission in the winter. The shading configuration can be customized to suit regional locations and seasons. Not only does the system reduce solar heat gain, but it also provides thermal insulation. SITumbra is composed of recyclable and renewable light-weight materials to offer structural strength, transparency, and energy efficiency. The system is manufactured to customized specifications for size, geometry, color, and materials.

CONTENTS
Recyclable polymers and biocomposite materials

APPLICATIONS
Large-scale window facade systems

TYPES / SIZES
Nominal panel size 4 x 8' (1.2 x 2.4 m) with spans up to 12' (3.7 m) for specialized applications; thickness 2–6" (5.1–15.2 cm); finishes: transparent, translucent, and customizable; variety of sizes, shapes, colors, and materials

ENVIRONMENTAL
Highly energy efficient (high R value, variable SHGC value), recyclable and renewable materials

TESTS / EXAMINATIONS
ASTM, ANSI, NFRC

LIMITATIONS
Not economical in small applications

CONTACT
Harry Giles
2232 South Main Street, #364
Ann Arbor, MI 48103
www.situmbra.com
info@situmbra.com

Solar Ivy

SOLAR ENERGY DEVICE INSPIRED BY IVY

Solar Ivy is a solar energy generation and delivery system inspired by ivy. Attached to a building facade, its "leaves" are flexible photovoltaic panels that flutter in the wind, creating a kinetic experience. Solar Ivy's visual appeal and flexible hardware system bring a technology traditionally restricted to the roof to almost any architectural surface.

Solar Ivy's proprietary design allows each leaf to be customized through color, angle, and rotation. Every leaf in a Solar Ivy installation is individually positioned to capture maximum energy from the sun. Solar Ivy's modular system of components gives users the option to upgrade or expand as solar technology progresses. The modular system ensures that Solar Ivy can be fully disassembled when repairs are necessary, and is completely recyclable at the end of its life cycle.

CONTENTS
Konarka Power Plastic, UV-stable high-density polyethylene plastic leaves, stainless-steel cable, X-tend mesh, stainless-steel hardware, copper core wire, stainless-steel clip hardware

APPLICATIONS
Solar power generation

TYPES / SIZES
Custom sizes and custom colors

ENVIRONMENTAL
Renewable energy production, recyclable/ reclaimable materials, efficient use of materials

TESTS / EXAMINATIONS
RoHS and UL testing are now underway

LIMITATIONS
A limited life span for organic photovoltaics (OPVs)—currently 5–10 years

CONTACT
SMIT
63 Flushing Avenue, Unit 195, Building 280, Suite 515
Brooklyn, NY 11205
Tel: 718-399-4452
www.s-m-i-t.com
contact@s-m-i-t.com

Solar Water Tarp

PERSONAL VESSEL FOR CARRYING AND DISINFECTING WATER

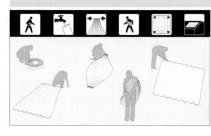   

The Solar Water Tarp is fabricated to be flexible and robust. Layers of rubberized nylon and high-performance LDPE are radio-frequency welded to produce durable, watertight cells. The cellular construction of the tarp, morphologically inspired by the saguaro cactus, is designed to conform to the body and varied volumes of water. Designed by Superficial Studio's Eric Olsen, the digitally designed pattern for the Solar Water Tarp lends itself to mass variation and is designed to be easily appropriated for a variety of situations: from carrying water to creating a sun shade; from urban rooftops to rural huts.

CONTENTS
Recycled low-density polyethylene (LDPE), coated nylon

APPLICATIONS
Water disinfection

TYPES / SIZES
71.3 x 39.4 x 2" (180 x 100 x 5 cm); 4.8 gal (18 l) capacity; custom forms possible

ENVIRONMENTAL
Recycled material, no power or chemicals required

CONTACT
Superficial Studio
3163 North Beachwood Drive
Hollywood, CA 90068
www.superficialstudio.com
info@superficialstudio.com

STRUCTURAL FOAMS

 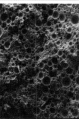  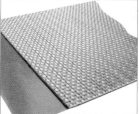

Henkel's series of Terocore foams consist of heat-activated or two-component epoxy foam products for structural applications such as reinforcement, durability, and crashworthiness. They are effectively being used in automotive and other industries to achieve competitive, lightweight, and high-performance products made from a wide variety of materials.

Terocore foams are applied onto metal or plastic carriers by extrusion or injection molding and provided as preformed parts to both the original equipment manufacturer (OEM) and sub-suppliers. The preformed parts can easily be integrated into the constructions of the customer. Henkel also provides full engineering and design support (e.g., FEA methods and data).

CONTENTS
Structural foams based on epoxy chemistry

APPLICATIONS
Improvement of stiffness, strength, durability, crashworthiness, and noise, vibration, and harshness (NVH) performance of automotive structures and constructions

TYPES / SIZES
Preformed parts tailored to customer demands including metal or plastic carriers

ENVIRONMENTAL
Reduced weight

TESTS / EXAMINATIONS
Preformed parts are designed and tested in accordance with customer requirements; meets automotive quality standards (e.g., ISO/TS 16949 certification)

LIMITATIONS
Foams will permanently bond to the substrates

CONTACT
Henkel AG & Co. KGaA
Henkel Teroson Strasse 57
Heidelberg, 69112
Germany
Tel: +49 6221-7040
www.teroson.com

Terrewalks

RECYCLED MODULAR INTERLOCKING PAVING TILES

Terrewalks is an alternative to concrete sidewalks, made of recycled tire rubber and reused PET plastic. Terrewalks is pervious and unbreakable, and suitable for all climates. The interlocking paving tiles assist groundwater recharge and heat island reduction, and provide a safe and comfortable walking surface. According to manufacturer Rubbersidewalks, one square foot of Terrewalks diverts 36 pounds (16.3 kilograms) of waste rubber and plastic from landfills, and each 20-square-foot (1.9-square-meter) installation saves one tree from removal. In addition, the low-energy manufacturing process results in a low carbon footprint.

CONTENTS
100% recycled crumb tire rubber, low-density polyethylene (PET) waste plastic with colorant

APPLICATIONS
Public and commercial sidewalks, walkways, plazas, courtyard, malls, connecting or garden walkways, office buildings, corporate campuses, high-rise residential development, retail complexes, hotels, casinos, medical centers, universities, sports venues, senior centers

TYPES / SIZES
24 x 30 x 1 7/8" (61 x 76.2 x 4.8 cm); colors: TerreGranite (gray), TerreCool (white), and more colors soon to be produced

ENVIRONMENTAL
100% recycled content, recyclable, directs water into soil, low-energy manufacturing process

TESTS / EXAMINATIONS
ASTM F355, C1028, D3884, D4762, B117, C1026, D6944, E662; Xenon Arc Weathering and Porosity DIN 18035

LIMITATIONS
Not for use in heavy vehicle-traffic areas

CONTACT
Rubbersidewalks, Inc.
10061 Talbert Avenue
Fountain Valley, CA 92708
Tel: 714-964-1400
www.rubbersidewalks.com
danjoyce@rubbersidewalks
.com

Plastic

HIGH-CONTOUR ARCHITECTURAL ACRYLIC

 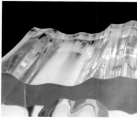 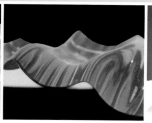 

R-Cast Textures acrylic material weighs half as much as glass but is seventeen times stronger. It is also four times stronger than concrete. Textures are acrylic panels that feature customized patterns sculpted deep into the surface. Without any special illumination, the panels reflect available ambient lighting to produce deep and dramatic shadows. Textures panels are available in opaque, clear, and translucent colors, and may be cut to custom shapes and sizes.

CONTENTS

Polymethyl methacrylate (PMMA)

APPLICATIONS

Decorative paneling, partitions, water features, signage, furniture, displays

TYPES / SIZES

Sheet size 4 x 8' (1.2 x 2.4 m); thickness 1" (2.5 cm); finishes: clear, frosted, opaque, custom colors

ENVIRONMENTAL

Manufactured in a facility that recycles waste and water to reduce our carbon footprint

TESTS / EXAMINATIONS

ASTM-D256, D570, D621, D638, D648, D695, D696, D732, D785, D790, D792

LIMITATIONS

Not available thinner than 3/4" (1.9 cm)

CONTACT

Reynolds Polymer Technology, Inc. 607 Hollingsworth Street Grand Junction, CO 81505 Tel: 800-433-9293 www.reynoldspolymer.com customerservice@ reynoldspolymer.com

MULTIDIMENSIONAL MATERIAL

Topiary Tile

FLEXIBLE LIGHT- AND SOUND-CUSHIONING MATERIAL

When the International Center of Photography approached Matter Practice to design an installation for presenting isolated audiovisual works for an exhibition, they devised a strategy for producing extremely lightweight, semirigid sound- and light-isolating panels requiring minimal structure for support. The panels can be assembled into a wide range of plastic forms—with easy field trimming and modification—to create rooms, screens, sculptures, seating, or wall surfacing.

While the material sourced for the initial test application was a particular density and color of polyethylene foam (typically used as plumbing pipe insulation), the panels can be fabricated out of any foam that can be sliced into rings. The shape of the aggregate and the direction and density in which it is pushed through the support netting causes the panels to have a slight outward bow, giving the panels structural rigidity and a voluminous expression.

CONTENTS
Polyethylene foam, nylon/
wire netting

APPLICATIONS
Audiovisual rooms, event
spaces, recreational spaces,
sound- and light-
dampening rooms

TYPES / SIZES
Made to order

ENVIRONMENTAL
100% recyclable

LIMITATIONS
Not suitable for outdoor
applications

CONTACT
Matter Practice
20 Jay Street, Suite 302
Brooklyn, NY 11201
Tel: 718-855-2255
www.matterpractice.net
matters@matterpractice.net

VapourGuard

WATER-HARVESTING FABRIC

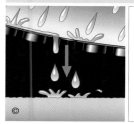   

VapourGuard is a dual-extruded polyethylene material used for dams and reservoirs to harvest and store water. Small drainage holes allow rainwater to be collected through the material, and the evaporation rate through the material is under 2 percent. The light top surface of VapourGuard reflects the sun's heat away from the water, while the dark underside prevents light from penetrating the surface, thus inhibiting algae growth in the water. The material has a weld edge to join sections of the cover together to create strong welded seams for covering expanses of water, and airborne debris is substantially reduced as a result.

CONTENTS
Dual-extruded polyethylene

APPLICATIONS
Floating cover for dams and reservoirs; stores water for use in irrigation, agriculture, hydroponics, mining, and drinking water

TYPES / SIZES
Installation area 6458–861,113 ft² (600–80,000 m²); width of 6.6' (2 m) or 8.2' (2.5 m); rolls 295' (90 m) long; 540 microns thick

ENVIRONMENTAL
100% recyclable, harvests rainfall, eliminates water evaporation by more than 98%

TESTS / EXAMINATIONS
Research carried out in collaboration with Brighton and London Metropolitan Universities proved that the cover eliminates water evaporation by more than 98% and inhibits algae growth

LIMITATIONS
Not suitable for swimming pool applications due to pool water treatment products; the VapourGuard Logo is registered by Plastipack Ltd. in the UK—it will soon be launched under a different name in the US

CONTACT
Plastipack Limited
Wainwright House,
4 Wainwright Close,
Churchfields Industrial Estate
St. Leonards-on-Sea,
East Sussex, TN38 9PP
United Kingdom
Tel: +44 (0)1424-851659
www.plastipack.co.uk
info@plastipack.co.uk

Varia Ecoresin

LAMINATED 40 PERCENT PRECONSUMER RECYCLED PETG

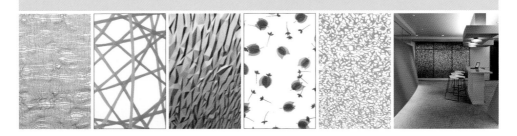

Varia Ecoresin is a dynamic interlayer system composed of partially preconsumer-recycled PETG. The system allows for the custom selection of colors, patterns, textures, print interlayers, and finishes in order to provide a vast number of design options. Varia Ecoresin has been engineered to incorporate 40 percent preconsumer grind content without compromising its overall physical properties. The material is also compatible with one of the largest postconsumer recycle streams, and is GREENGUARD Level 3 Indoor Air Quality–Certified.

CONTENTS
Polyethylene terephthalate glycol modified (PETG), organic, textile, print interlayers

APPLICATIONS
Vertical and horizontal surfaces, ceilings, furniture, lighting, signage, acoustics

TYPES / SIZES
4 x 8' (1.2 x 2.4 m), 4 x 10' (1.2 x 3 m); custom sizes available, multiple finishes available

ENVIRONMENTAL
GREENGUARD-certified for indoor air quality, SCS certified 40% preconsumer recycled content

TESTS / EXAMINATIONS
ASTM E84: 1/4" to 3/4" gauges=CLASS B, 1" gauge=CLASS A; NFPA 286: 1/4" to 3/8" gauges= CLASS A, PASS CC1 for Light Transmitting Plastics; ANSI Z87.1 (Safety Glazing)—PASS

LIMITATIONS
Maximum temperature use 150°F (65.5°C); requires UV protection and edge sealing for exterior use

CONTACT
3form
2300 South 2300 West
Salt Lake City, UT 84119
Tel: 801-649-2500
www.3-form.com
info@3-form.com

NATURAL GROWTH-PATTERN CHAIR

While at work developing their notable Algues product, Ronan and Erwan Bouroullec began to speculate whether similar principles of branching and growth could be applied to a chair design. Based on precedents found in the works of Axel Erlandson, Alvar Aalto, Michael Thonet, and others, the Bouroullec brothers began to conceptualize a chair composed of four branches that grow into a network of twigs functioning as a basketlike seat. Unlike the precedents, who worked with natural materials, the Bouroullec brothers wanted to fabricate this chair out of plastic—relishing the inherent contradiction between natural growth processes and industrial mass production. The Vegetal Chair reveals the designers' interest in harnessing nature's principles of growth in order to create a kind of artificial nature.

CONTENTS
Fiber-reinforced polyamide

APPLICATIONS
Seating

CONTACT
Vitra
Klünenfeldstrasse 22
Birsfelden, CH-4127
Switzerland
Tel: +41 61-3770000
www.vitra.com

Watercolors

HIGH-CONTRAST INTERACTIVE FLOORING

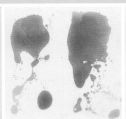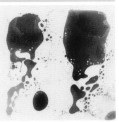

B-surfaces Watercolors, designed by Gianfranco Barban and Gregg Brodarick, are modular floor tiles that contain multiple fluids encapsulated between sealed polymer sheets. Like B.lab Italia's other liquid-filled surfaces, Watercolors respond to physical touch and may be used on various horizontal surfaces to invite playful interaction. Unlike their other surfaces, however, Watercolors tiles contain a combination of opaque and translucent nonmixing fluids. This pairing results in greater contrast and clarity of visual pattern.

CONTENTS
Polycarbonate (PC),
polyvinyl chloride (PVC),
polyethylene terephthalate
(PET), nontoxic liquids

APPLICATIONS
Flooring for residential and
commercial interiors

TYPES / SIZES
Tiles 39 x 39" (99 x 99 cm)
or 19 1/2 x 19 1/2" (49.5 x
49.5 cm)

ENVIRONMENTAL
Nontoxic fluids

TESTS / EXAMINATIONS
Certified flooring product
for residential and
commercial interiors, Class
1 fire rating

LIMITATIONS
For interior use only

CONTACT
B.lab Italia
Via Marmolada 20
Gallarate, VA 21052
Italy
Tel: +39 0331-774445
www.blabitalia.com
bmail@blabitalia.com

Zcell

CROSS-LINKED EVA FOAM WITH ENGINEERED CELL STRUCTURES

Zcell is an engineered foaming cell-structure technology that provides customizable material solutions related to protection requirements. Developed by South Korea-based DXD Inc., Zcell technology offers a broad range of lightweight cushioning and protective benefits with improved structure, performance, function, flexibility, and consistency in cross-linked foam components and products. Zcell allows for unlimited cell structure patterns, multiple molding processes, multiple materials and combinations, as well as unlimited configurations.

CONTENTS
Ethyl vinyl acetate (EVA), polyurethane (PU), blown rubber; or a combination of any of these materials, including Gel and TPU, as second fillers of cell structures

APPLICATIONS
High-demand applications for cushioning or protection including footwear, sporting equipment, electronics, automotive, building insulation and acoustics, medical, and marine applications

TYPES / SIZES
Custom sizing based on structural design

ENVIRONMENTAL
Waste reduction, elimination of VOCs, no lamination required

TESTS / EXAMINATIONS
KIFLT (Korea Institute of Footwear & Leather Technology) testing

LIMITATIONS
Piece production design only (no continuous roll)

CONTACT
DXD Inc.
Hwasin Apartment 1-401,
Milak-dong, Suyoung-gu
Busan, 613-828
South Korea
Tel: +82 51-756-3906
www.edxd.com
nextdot09@gmail.com

06: **GLASS**

24K Blown Glass

HAND-BLOWN GLASS INFUSED WITH 24-KARAT GOLD

Suzan Etkin Enterprises combines refined metals with glass to create exotic material hybrids. Gold is fused to the glass in its molten state, allowed to cool, and then re-covered with another layer of glass so that the gold is embedded in the material—creating objects with an internal glow.

24K Blown Glass vessels can be internally coated with clear, flexible, and adhesive antishatter resin, and glass elements can become structural when reinforced with steel embedded in clear resin. Each project is engineered and tested for safety.

Suzan Etkin Enterprises specializes in unique collaborative ventures with architects and interior designers. The design process requires exploration, engineering, and testing to coax the inherently organic material into precise and safely designed structures while maintaining the inherent character of blown glass.

CONTENTS
Glass, 24-karat gold

APPLICATIONS
Art features, lighting, feature walls, custom objects

TYPES / SIZES
Custom (other precious metals available)

ENVIRONMENTAL
100% recyclable

TESTS / EXAMINATIONS
Testing performed for each project

CONTACT
Suzan Etkin Enterprises
63 Greene Street, Suite 404
New York, NY 10012
Tel: 212-431-4176
www.suzanetkin.com
info@suzanetkin
enterprises.com

GEOMETRIC CAST GLASS

Exploiting the potential for enhanced dimensionality in glass, Nathan Allan has developed new production methods for tempering and laminating their Convex Series. Developed in partnership with Janson Goldstein and Front Inc. of New York City, this new glass may be profiled with various geometric patterns to create an undulating, multifaceted surface. Convex Series glass can be produced as single-layered panels from 1/4 inches to 3/4 inches (.6 centimeters to 1.9 centimeters) thick, and can be safety tempered as well. It is available in clear and low-iron glass. Cast textures and privacy coatings are also available.

CONTENTS
Glass with up to 30% recycled content

APPLICATIONS
Facades, partitions, feature walls

TYPES / SIZES
Patterns: circle, triangle, square, custom; finishes: clear, firefrost; sizes: 1 x 1' (.3 x .3 m) to 7 x 12' (2.1 x 3.7 m); thicknesses: 1/4", 5/16", 3/8", 1/2", 5/8", 3/4" (.6 cm, .8 cm, .95 cm, 1.3 cm, 1.6 cm, 1.9 cm)

ENVIRONMENTAL
Up to 30% recycled cullet, recyclable, efficient use of material

TESTS / EXAMINATIONS
Tempered, laminated (polyurethane resin)

LIMITATIONS
Maximum size 7 x 12' (2.1 x 3.7 m)

CONTACT
Nathan Allan Glass Studios Inc.
110-12011 Riverside Way
Richmond, BC V6W 1K6
Canada
Tel: 604-277-8533 x225
www.nathanallan.com
bm@nathanallan.com

MULTIDIMENSIONAL MATERIAL

Crackle

SCULPTURAL CAST GLASS

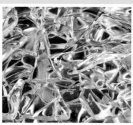

When French luxury jeweler Mauboussin approached the Rockwell Group to design its new store, the design firm selected Nathan Allan Glass Studios Inc. to create a unique architectural decor element. The challenge was to make a statement without overwhelming the high-end merchandise it would showcase. Working in conjunction with Rockwell's designers, Nathan Allan developed an icy, glittering new glass product that echoes the chiseled surfaces of brilliant diamonds. Entitled Crackle, this glass was used to unify three floors while complementing Mauboussin's exquisite pieces.

CONTENTS
Glass with up to 30% recycled content

APPLICATIONS
Cladding, partitions, feature walls

TYPES / SIZES
Types: clear, low iron; maximum size 4 x 4' (1.2 x 1.2 m); overall thickness 1" (2.5 cm)

ENVIRONMENTAL
Up to 30% recycled cullet, recyclable, efficient use of material

TESTS / EXAMINATIONS
Laminated (polyurethane resin)

LIMITATIONS
Maximum size 4 x 4' (1.2 x 1.2 m)

CONTACT
Nathan Allan Glass Studios Inc.
110-12011 Riverside Way
Richmond, BC V6W 1K6
Canada
Tel: 604-277-8533 x225
www.nathanallan.com
bm@nathanallan.com

Crystallized Glass Stone

GLASS-STONE TILES AND SLABS

Crystallized Glass Stone tiles and slabs are engineered from glass materials to create a high-density, heat- and wear-resistant product for wall surfaces, flooring, and countertops. The smooth, nonporous surface is luminous and subtly reflective, creating a monolithic aesthetic as a creative alternative to stone. Crystallized Glass Stone is available in a range of colors and patterns.

CONTENTS
Glass

APPLICATIONS
Flooring and surfacing applications in all commercial environments

TYPES / SIZES
24 x 24 x 1/2" (61 x 61 x 1.3 cm); slabs 24 x 24 x 3/4" (61 x 61 x 1.9 cm); custom sizes available

ENVIRONMENTAL
Replacement for stone

LIMITATIONS
Interior use only

CONTACT
Architectural Systems, Inc.
150 West 25th Street,
8th Floor
New York, NY 10001
Tel: 800-793-0224
www.archsystems.com
sales@archsystems.com

Crystalmeshglass

TEXTURED, REFLECTIVE TILING SYSTEM

Crystalmeshglass is an enhanced, flexible tiling system composed of crystals, stained glass, and mirror pieces connected to a fiberglass mesh backing that can be easily overlaid onto complex, curvilinear surfaces. Crystalmeshglass interior and exterior decor surfacing bonds to most surfaces, including ceilings, walls, columns, counters, architectural accents, and murals. Moreover, the glass used is UV-stable, 100 percent recyclable, and the colors are permanent.

Created by Duncan M. Kirk, Crystalmeshglass surfaces may be designed and customized using software available online at www.meshglass.com, which issues specific codes used in the manufacturing of each design.

CONTENTS
Swarovski crystals, sheet mirror, sheet glass, fiberglass mesh, 3M Fastbond adhesive

APPLICATIONS
Decor surfacing for interiors, exteriors, and objects

TYPES / SIZES
Tiles 12 x 12" (30.5 x 30.5 cm); border trims 12 x 4" (30.5 x 10 cm); weight 1.5 lbs/ft^2 (7.3 kg/m^2)

ENVIRONMENTAL
Glass is 100% recyclable

LIMITATIONS
Not for use as flooring

CONTACT
Meshglass Ltd.
299 East 10th Street, #5
New York, NY 10009
Tel: 636-486-4410
www.meshglass.com
info@meshglass.com

Fusion

TEXTURED DICHROIC CAST GLASS

Fusion incorporates the iridescent color features of dichroic glass with clear or low-iron options. Fissures and air bubbles also become integral components of the finished glass pieces, making the final appearance even more visually captivating. Composed of up to 30 percent recycled content and offered in multiple colors and finishes, Fusion is a celebration of the robust decorative potential of glass.

CONTENTS
Glass with up to 30% recycled content

APPLICATIONS
Cladding, partitions, feature walls

TYPES / SIZES
Maximum size 30 x 72" (76.2 x 182.9 cm); thickness 1/2" (1.3 cm) or 2" (5.1 cm); colors: Beamz, Breez, Flex

ENVIRONMENTAL
Up to 30% recycled cullet, recyclable, efficient use of material

TESTS / EXAMINATIONS
Tempered, laminated (polyurethane resin)

LIMITATIONS
Air bubbles and fissures may form in thicker material

CONTACT
Nathan Allan Glass Studios Inc.
110-12011 Riverside Way
Richmond, BC V6W 1K6
Canada
Tel: 604-277-8533 x225
www.nathanallan.com
bm@nathanallan.com

G Series

RECYCLED GLASS ARCHITECTURAL COATING

EverGreene's G Series glass-coating system is composed of 100 percent recycled glass, suspended in a natural binder, then hand-applied to achieve a jewellike, luminescent surface. This environmentally friendly material is semitranslucent and visually dynamic. Because it is not constrained by strips or paneling, it can be applied to an unrestricted, continuous surface. The size of the glass beads used is customizable, and the coating has a subtle sheen that gently reflects and refracts both natural and artificial light.

CONTENTS
100% postconsumer recycled glass

APPLICATIONS
Wall and ceiling application, customized canvas application

TYPES / SIZES
Custom

ENVIRONMENTAL
Low-VOC paints and sealants

LIMITATIONS
Not for exterior use

CONTACT
EverGreene Architectural Arts, Inc.
450 West 31st Street,
7th Floor
New York, NY 10001
Tel: 212-244-2800
www.evergreene.com
info@evergreene.com

GreenPix

ZERO ENERGY MEDIA WALL

GreenPix is the first zero energy media wall, absorbing solar energy during the day and then powering the media wall at night. The panels can be used to create stunning media effects on very large building envelopes that are viewable from both inside and outside the building.

Designed by Simone Giostra, GreenPix is a transparent media wall for dynamic content display, including playback videos, interactive performances, and live and user-generated content. Its "intelligent skin" interacts with building interiors and outer public spaces using embedded, custom-designed software, transforming the building facade into a responsive environment for entertainment and public engagement. GreenPix allows daylight into the building while reducing its exposure to direct sunlight. The photovoltaic density pattern increases the building's performance, allowing natural light when required by interior program, while reducing heat gain and transforming excessive solar radiation into energy for the media wall. The photovoltaic system can be connected to local battery storage or to the grid, and power can be sold back to the electricity company.

CONTENTS
Low-resolution LED lighting, photovoltaic cells, laminated glass

APPLICATIONS
Solar-powered media wall for low-resolution, large-scale installations

TYPES / SIZES
Low resolution, no size limitation

ENVIRONMENTAL
Reduces heat gain, uses renewable energy, low-energy integral lighting

TESTS / EXAMINATIONS
UL certified, patent pending

LIMITATIONS
Ideal for low-resolution artistic media content

CONTACT
GreenPix LLC
55 Washington Street, #454
Brooklyn, NY 11201
Tel: 212-920-8180
www.greenpix.org
info@greenpix.org

KiloLux

GLASS BRICK

KiloLux is a monolithic glass block with the appearance of a laminate, produced by thermally bonding glass at low temperatures. Wales-based Innovative Glass Products developed the material for inclusion in building facades as a light-emitting decorative element. Because no adhesive is used to manufacture the product, it has good resistance to moisture and freeze-thaw cycles. KiloLux was first employed as an external component and weather seal on a public building in the United Kingdom, and had to be rigorously tested to ensure compliance with building specifications. Since then, it has been used in a variety of interior or exterior applications and the product shape and color offerings have been diversified. KiloLux can be manufactured with a variety of different glass types, most commonly with soda-lime float glass and low-iron glass. It can be secondarily processed using a variety of chemical and mechanical abrasive techniques, and can also be thermally bent as well as cut and shaped by a diamond saw.

CONTENTS
Soda lime glass

APPLICATIONS
Interior and exterior applications

TYPES / SIZES
12 x 4 x 2" (30.5 x 10.2 x 5.1 cm), custom sizes available, including typical brick dimensions

ENVIRONMENTAL
100% recyclable

TESTS / EXAMINATIONS
Tested for freeze-thaw, solar shading, and manufactured to edge stress tolerances using photoelastic stress measurements

LIMITATIONS
Size limitations

CONTACT
Innovative Glass Products
Unit 9C, Players Industrial Estate, Clydach
Swansea, Wales SA6 5BQ
United Kingdom
Tel: +44(0) 77968-581477
www.innovativeglass.co.uk
enquiry@innovativeglass
.co.uk

HAND-BLOWN SILVERED GLASS

Silvered-glass "mercury glass" was invented in the nineteenth century as a decorative substitute for more expensive silver tableware. Where traditional mercury glass has a thin glass wall, however, Suzan Etkin Enterprises creates hand-blown silvered glass with thick, undulating walls and a film of pure silver. The thick, wavy walls give the glass the appearance of liquid metal, transforming a nineteenth-century technology into a twenty-first-century material with unmatched light reflection and refraction properties.

Suzan Etkin engineers and tests each project for safety. Overhead glass vessels can be filled with antishatter foam or resin, and mercury glass elements can assume structural properties when reinforced with steel embedded in structural foam or resin.

CONTENTS
Glass, pure silver, lacquer

APPLICATIONS
Art features, lighting, feature walls, custom objects

TYPES / SIZES
Custom

ENVIRONMENTAL
100% recyclable

CONTACT
Suzan Etkin Enterprises
63 Greene Street, Suite 404
New York, NY 10012
Tel: 212-431-4176
www.suzanetkin.com
info@suzanetkin
enterprises.com

RECOMBINANT MATERIAL

Northern Lights

THERMOCHROMIC GLASS

Northern Lights, designed by Blane Kivley, is a glass product that can change its appearance with a change in temperature. Human touch, an adjustment in the ambient air temperature, hot or cold water, or any relatively warm or cool source will trigger a thermochromic response. Northern Lights material may be designed in any size, color, or thickness of glass, as well as with multiple activation temperatures. Moreover, the material's base color may be matched to custom colors.

CONTENTS
Glass

APPLICATIONS
Tile, glass blocks, lighting, glass vessels, shower walls, architectural glass, countertops, tables, floors, medallions, ceilings, roof tiles, walls, doors

TYPES / SIZES
4 x 4 x 3/8" (10.2 x 10.2 x .95 cm), custom sizes available

ENVIRONMENTAL
Up to 20% recycled glass

LIMITATIONS
Handmade version for interior use only

CONTACT
Moving Color
2351 Sunset Boulevard, Suite 425
Rocklin, CA 95765
Tel: 916-337-6296
www.movingcolor.net
blane@movingcolor.net

CAST RESIN WITH ENCAPSULATED INTERLAYERS BETWEEN GLASS LITES

   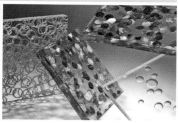

3form Poured Glass is produced by encapsulating organic materials and fabrics with a poured cast resin. This technique enables the possibility of conveying depth without sacrificing clarity or the natural beauty of the interlayer materials. Poured Glass is based on a proprietary liquid-lamination technology that sets itself apart from conventional methods in two ways: first, the lamination process allows even distribution of objects throughout the glass; and second, it makes possible clean, exposed edges that allow for a frameless application. The result is a seamless fusion of glass, resin, and interlayers into visually dramatic panels whose interlayers are delicately and dimensionally preserved.

CONTENTS
Glass, poured thermoset resin, textiles, organics, and botanicals

APPLICATIONS
Interior applications including horizontal surfaces and feature walls

TYPES / SIZES
Lengths 8' (2.4 m), 10' (3 m); widths 3' (.9 m), 4' (1.2 m); gauge 3/4"

TESTS / EXAMINATIONS
Safety Glazing ANSI Z97.1

LIMITATIONS
No heat forming, cold forming, or exterior applications

CONTACT
3form
2300 South 2300 West
Salt Lake City, UT 84119
Tel: 801-649-2500
www.3-form.com
info@3-form.com

Pressed Glass

LAMINATED GLASS WITH INTERLAYERS

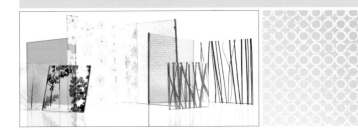
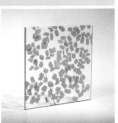

Pressed Glass offers the optical clarity and excellent fire performance of glass with a wide variety of interlayer options. 3form offers organic, printed, and textile materials for insertion between glass layers, and the result is an extremely rigid, thin-gauge product ideal for applications requiring a clean appearance and minimal hardware.

CONTENTS
Glass; organic, printed, and textile interlayers

APPLICATIONS
Feature walls, horizontal and vertical surfaces, privacy windows

TYPES / SIZES
Maximum length 10' (3 m); maximum width 4' (1.2 m); custom sizes available; gauges: 5/16", 3/8", 7/16", 1/2", 9/16", 5/8", 11/16", 3/4", 13/16", 7/8", 1", 1 1/16", 1 1/8", 1 3/16", 1 1/4", 1 5/16" (8 mm, 9.5 mm, 11 mm, 13 mm, 14 mm, 16 mm, 17 mm, 19 mm, 21 mm, 22 mm, 25 mm, 27 mm, 29 mm, 30 mm, 32 mm, and 33 mm)

TESTS / EXAMINATIONS
ANSI Z97.1 (Safety Glazing)

LIMITATIONS
No heat forming or cold bending

CONTACT
3form
2300 South 2300 West
Salt Lake City, UT 84119
Tel: 801-649-2500
www.3-form.com
info@3-form.com

RECYCLED PRECONSUMER LEAD CRYSTAL GLASS

 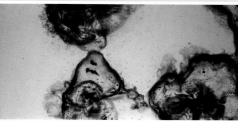

Fernando Miguel Marques collects postindustrial waste glass from Atlantis, the largest Portugese producer of lead crystal glass. Marques has developed a technique that allows the repurposing of the glass into tiles, panels, and other shapes. Recycled Crystal Glass may be used for architectural surfaces as well as decorative objects, and the fabrication process allows for partial staining or total vitreous-mass staining.

CONTENTS
100% recycled glass

APPLICATIONS
Interior and exterior tiles for wall surfaces, nonfood-based decorative or utilitarian products

TYPES / SIZES
Tiles 5.5 x 3.5 x .4"
(14 x 9 x 1 cm)

ENVIRONMENTAL
100% preconsumer recycled content

LIMITATIONS
Nonfood products

CONTACT
Fernando Miguel Marques
Estrada do Zambujal,
N°76, 2°A
Amadora, Lisboa 2610-193
Portugal
home.fa.utl.pt/~ciaud/fmm
.design@gmail.com

REPURPOSED MATERIAL

SentryGlas Expressions

EXTERIOR-GRADE, DIGITALLY PRINTED LAMINATED GLASS

Pulp Studio's SentryGlas Expressions is a computer-controlled digital imaging system for decorative glass, made with a UV-resistant ink that will not fade in the sun like other ink-jet technologies. Unlike ceramic frit, SentryGlas Expressions provides many more color and image-resolution options. Solid colors remain consistent, and gradient tones hold just as in the original digital image. SentryGlas Expressions allows for the creation of a wide range of transparency levels within a single laminated pane. It is possible to have solid colors or high-density photography in one part of the image and complete clarity elsewhere. It is also possible to screen images for an ephemeral quality. Pulp Studio can generate art proofs in a few days and when alterations are needed, image content can be modified quickly.

CONTENTS
Glass, ink-jet printed patented polyvinyl butyral (PVB) layer

APPLICATIONS
Curtain walls, facade or storefront glass, office partitions, room dividers, retail windows/counters, balustrades, shower doors, elevator interiors, conference room walls, advertising/signage, flooring

TYPES / SIZES
Any size regular or Low-E laminated piece of glass up to 65 x 120" (165 x 305 cm)

ENVIRONMENTAL
Daylighting applications

TESTS / EXAMINATIONS
Meets ANSI Z97 specifications for safety glass

CONTACT
Robin Reigi Inc.
48 West 21st Street,
Suite 1002
New York, NY 10010
Tel: 212-924-5558
www.robin-reigi.com
info@robin-reigi.com

LAYERED CAST GLASS

Named after the way each piece is stacked then fused together, Stax series offers a way to create rich, multilayered textures using glass. Stax is designed to be deployed in exterior feature walls over very large surfaces. Layers of glass are fused together to create solid panels, and the face surface of each panel is staggered.

Three patterns are currently available: Beamz (straight planks), Breez (gentle rises), and Flex (twisting fluidity). Although Stax is too thick to temper, flat panels of clear mirror glass may be laminated to the back of the panels in order to classify as safety glass.

CONTENTS
Glass with up to 30% recycled content

APPLICATIONS
Cladding, partitions, feature walls

TYPES / SIZES
Panels up to 4 x 9' (1.2 x 2.7 m); typical thickness 1 1/4" (3.2 cm), may vary from 1–1 1/2" (2.5–3.8 cm); patterns: Beamz, Breez, Flex; finishes: clear, amber with mirror

ENVIRONMENTAL
Up to 30% recycled cullet, recyclable

TESTS / EXAMINATIONS
Tempered, laminated (polyurethane resin)

LIMITATIONS
Too thick to temper

CONTACT
Nathan Allan Glass Studios Inc.
110-12011 Riverside Way
Richmond, BC V6W 1K6
Canada
Tel: 604-277-8533 x225
www.nathanallan.com
bm@nathanallan.com

07: **PAINT + PAPER**

Catalyst

METALLIC HIGH-PRESSURE LAMINATE

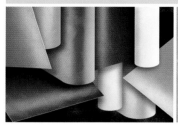

Lamin-Art's latest innovation transforms common high-pressure laminate (HPL)into a versatile and durable decorative surface with a notable luster. The Catalyst collection features a hand-crafted, spiral-etched dimensional finish that recreates the brilliance and texture of brushed metal. Unlike real metallic surfaces, however, Catalyst boasts the scratch- and dent-resistance properties of high-pressure decorative laminate, making the surface virtually maintenance-free. Catalyst's broad metallic color palette is suitable for a wide variety of commercial design concepts.

Catalyst meets NEMA standards for abrasion and scratch resistance, making it suitable for use on both horizontal and vertical surfaces requiring high visual impact and enhanced durability. In addition, the Catalyst collection lays-up and fabricates the very same way as any other high-pressure decorative laminate, and won't wear tools or saw blades.

CONTENTS
75% paper, 25% resin

APPLICATIONS
Unlike other textural laminates in the marketplace, Catalyst can be used as a decorative surface on both horizontal and vertical applications that require superior wear, impact, and stain resistance.

TYPES / SIZES
Sheet size 4 x 10' (1.2 x 3 m); thickness .05" (1.2 mm) or .03" (.7 mm)

ENVIRONMENTAL
Made from 22% preconsumer recycled content, sourced from responsibly managed forests that are Sustainable Forestry Initiative-certified, GREENGUARD certified

TESTS / EXAMINATIONS
NEMA Pub. LD3-2005, Section 2

LIMITATIONS
Not for exterior use; not recommended for direct application to plaster or gypsum wallboard, or to concrete

CONTACT
Lamin-Art
1670 Basswood Road
Schaumburg, IL 60173
Tel: 800-323-7624
www.laminart.com
info@laminart.com

Helicone HC

LIQUID-CRYSTAL EFFECT PIGMENTS

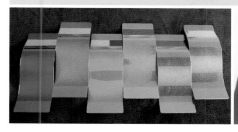  

Unlike commonly known inorganic-effect pigment technologies such as interference and metallics, Helicone HC is based on a purely organic polymer. It is made from customized fine chemicals that are then transferred into a liquid-crystal polymer film, and subsequently milled into an effect pigment flake. Helicone HC effect pigments are transparent platelets that exhibit no color on their own, but when used over, or in combination with, other more common colorants, exhibit a novel color transition effect. They also provide extreme visual depth and a deep metallic-like sparkle, resulting in a truly exceptional three-dimensional appearance. The polymeric platelets are highly versatile and show a superior dispersability in either water- or solvent-based coatings. Their low specific gravity leads to minimal settling. Furthermore, their elasticity and exceptional shear force resistance allow the Helicone HC effect pigments to be easily formulated into the melt mix in powder coatings, or extruded into plastics. They are nontoxic and contain no heavy metals.

CONTENTS
100% polymer

APPLICATIONS
Surface decoration

TYPES / SIZES
6 colors in 5 particle sizes

ENVIRONMENTAL
Nonhazardous polymeric
raw material

LIMITATIONS
Needs a protective top coat
for high performance
outdoor applications (e.g.,
automotive coatings)

CONTACT
LCP Technology GmbH
Johannes Hess Strasse 24
Burghausen, 84489
Germany
Tel: +1 989-6711084
www.helicone.com
info@lcptechnology.com

Premium Parchment

ARCHITECTURAL WASHI PARCHMENT

Precious Pieces harnesses a half-millenium tradition of making mulberry-fiber parchment in order to provide large, exquisitely crafted pieces of the highest quality Japanese washi. The Premium Parchment series contains thirty-seven standard pattern variations, with the option of designing custom pieces, available in seamless sheets as large as 30 x 100 feet (9.1 x 30.5 m). With this series, Precious Pieces aims to embody a skillful blend of contemporary design and traditional craft.

CONTENTS
100% Japanese mulberry, mitsumata, and/or ganpi fibers (washi parchment)

APPLICATIONS
Illuminated walls, tapestries, wall coverings, doors, furniture, lighting fixtures, lamp shades

TYPES / SIZES
37 pattern variations; custom seamless pieces up to 30 x 100' (9.1 x 30.5 m)

ENVIRONMENTAL
100% natural material, 100% handmade, no chemical utilization, zero emissions

LIMITATIONS
Not for exterior use without specific treatment

CONTACT
Precious Pieces
5 Tudor City Place, #102
New York, NY 10017
Tel: 212-682-8505
www.precious-piece.com
info@precious-piece.com

PAPER ELECTRONICS

 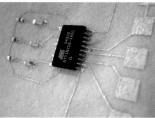 

Pulp-Based Computing, developed by Marcelo Coelho, is a fabrication technique for creating paper composites that can function as sensors, actuators, and circuit boards while retaining the physical and aesthetic qualities of paper. Papermaking allows for an inclusion process, where a physical object can be permanently embedded between two individual paper sheets that are then compressed, drained, and set to dry. By silk-screening and encapsulating electrically active inks, conductive threads, and smart materials between sheets, it is possible to create an electronic paper "sandwich" that is resilient and inseparable from its embedded object. This process allows for the fabrication of paper speakers, emissive displays, as well as bend-and-touch sensors. While electronic paper technologies usually overlook the material qualities that are at the core of paper's versatility, Pulp-Based Computing produces electronic paper composites that can be folded, shredded, recycled, stapled, and written on while preserving the electrical reliability and resilience of traditional electronic components.

CONTENTS
Cotton fiber paper, conductive and resistive inks, smart materials

APPLICATIONS
Light-emitting paper, bend sensors, paper speakers, multilayered circuit-board substrate, flexible electronics

TYPES / SIZES
A4 standard-size paper; custom dimensions and shapes possible

ENVIRONMENTAL
Natural fibers, recyclable

LIMITATIONS
Flammable; suitable for low-power applications only

CONTACT
MIT Media Laboratory
20 Ames Street, E15-322
Cambridge, MA 02139
Tel: 617-452-5695
www.cmarcelo.com
marcelo@media.mit.edu

SOUND Wall

SOUND RESPONSIVE WALL MODULES

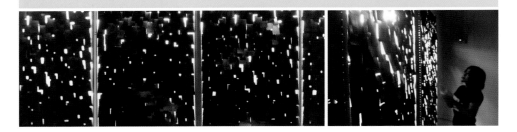

The NONdesigns SOUND Wall, designed by Scott Franklin and Miao Miao, is a modular system of interactive tiles used to create wall surfaces that alter their texture and light qualities based on environmental stimuli. With every sound produced by its audience, the SOUND Wall responds by opening its "skin," allowing light to spill out across the textural surface. Each module reacts independently, causing the wall to have reflections of activity where it is closest to sources of sound. The system is expandable to any wall size and its sensitivity is adjustable to react to soft or loud sound levels. The modules are custom-built for each project, so size and color can be selected to fit the application.

CONTENTS
Fan, microphone, Tyvek, birch plywood

APPLICATIONS
Interactive wall surface

TYPES / SIZES
Modules are available as 2 x 2' (.6 x .6 m) squares, which tile together to create any size wall covering

ENVIRONMENTAL
Refurbishable materials

LIMITATIONS
For indoor use only

CONTACT
NONdesigns, LLC
620 Moulton Avenue, #112
Los Angeles, CA 90031
Tel: 323-222-7883
www.nondesigns.com
info@nondesigns.com

PRINTING WITH BACTERIA

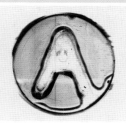

A course at the Department of Microbiology at the University of Wageningen taught Jelte van Abbema how to handle bacteria and their surroundings. Armed with this new knowledge and with the help of a microscope, Abbema began experiments in living font manipulation. According to Abbema, "the page becomes a feeding ground where bacteria can profilate, but it is the surrounding environment that determines the image's growth potential."

In recounting his experiences, the designer explains that "For these hungry little creatures to grow they require a nice hot and humid environment and food. Give them this and they basically grow on anything. In order to get the bacteria interested in the paper I used agar (a substance scientists use to cultivate bacteria in the laboratory). Fortunately some of them liked it, and some even started eating the cellulose off the paper. However if you let them grow without rules it immediately becomes chaotic. So to form images I had to control the shape of the bacterial culture right from the outset. For this I used various techniques such as silk-screen printing and old wooden-cut letters. At first the ink on the paper is hardly visible because the quantity of bacteria is minimal. But then, as they start to grow their pigment is unveiled and you begin to see them. In a converted poster box where the paper can reveal its life, messages appear and change through time."

CONTENTS
Growth medium, bacteria

APPLICATIONS
Living fonts, environmental signage, time-based graphics

TYPES / SIZES
Custom

CONTACT
Jelte van Abbema
Marcantilaan 351
Amsterdam, 10 51 NJ
The Netherlands
Tel: +31 640780800
www.vanabbema.net
jelte@vanabbema.net

Tatami Igusa

IGUSA STRAW WALL COVERING

    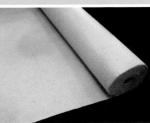

Precious Pieces has developed a new washi wall covering using igusa straw cultivated in Kumamoto, Japan. Igusa straw is the primary ingredient used to make tatami mats in Japan. The material has a variety of benefits such as air purification, mold and bacteria resistance, and humidity conditioning. The material also has a nice tactile quality and emits a delicate fragrance into interior spaces. Tatami Igusa imparts an uncanny warmth to interior wall surfaces, and is produced with no added chemicals.

CONTENTS
Tatami igusa straw, wood fiber

APPLICATIONS
Wall coverings

TYPES / SIZES
Rolls 39" (99 cm) wide, 150' (45.7 m) long; colors: plain, beige, charcoal

ENVIRONMENTAL
Air purification, zero emissions during manufacture, no chemical usage

LIMITATIONS
Not for exterior use

CONTACT
Precious Pieces
5 Tudor City Place, #102
New York, NY 10017
Tel: 212-682-8505
www.precious-piece.com
info@precious-piece.com

LAMINATED GLASS WITH WASHI PARCHMENT AND PVB

 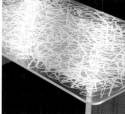   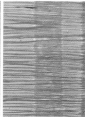

RECOMBINANT PRODUCT

Washi Laminated Glass panels are a combination of two glass sheets with handmade Japanese washi parchment and one interlayer of plastic (PVB). The hybrid material is used for both security and decorative applications, and demonstrates the marriage of a five-hundred-year-old, handcraft-based tradition with modern technology.

CONTENTS
Washi parchment made of 100% Japanese mulberry, ganpi, and/or mitsumata fibers, glass, polyvinyl butyral (PVB)

APPLICATIONS
Illuminated wall, space divider, doors, windows, furniture, lighting fixtures

TYPES / SIZES
Maximum size 5 x 10' (1.5 x 3 m); thickness 1/8–1" (3.2–25 mm); full line of washi patterns available

CONTACT
Precious Pieces
5 Tudor City Place, #102
New York, NY 10017
Tel: 212-682-8505
www.precious-piece.com
info@precious-piece.com

Wellboard

CORRUGATED CELLULOSE-SHEET MATERIAL

  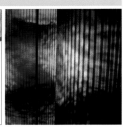

Wellboard consists of 100 percent cellulose with no adhesives or binders. Different profiles are pressed into a flat baseplate using heat and pressure. In spite of its high stability, Wellboard is a lightweight material, and its corrugated surface and flexibility make it an excellent material for exhibition, retail, and furniture design. It can be handled like wood and can be sawn, drilled, sanded, or glued. The surface of Wellboard can be given various colors using varnishes and glazes, and it can also be three-dimensionally printed.

CONTENTS
100% cellulose (no added adhesives or binders)

APPLICATIONS
Furniture creation, shop construction, interior design, exhibition construction, product design

TYPES / SIZES
Corrugated profile, gamma with trapeze-shaped profile

ENVIRONMENTAL
100% recyclable, highly efficient

TESTS / EXAMINATIONS
Fire protection class B2 (can be enhanced to B1 via impregnation with flame retardant)

LIMITATIONS
Not for exterior use, avoid humid environments

CONTACT
Well Ausstellungssystem GmbH
Schwarzer Bär 2
Hannover, D-30449
Germany
Tel: +49 511-92881-10
www.well.de
info@well.de

08: **FABRIC**

Choreographed Geometry

CONSTRUCTED THREE-DIMENSIONAL FABRIC

Choreographed Geometry is a systematically developed textile object whose geometry is exposed to a number of manipulative processes through its use in connection with the human body. It is the further development of an ongoing investigation into geometric organization, textile materiality, and its resulting transformative potential. Differently folded geometrical modules relate in size, scale, and geometry to certain body extremities and can be animated by the user when connected to a relational textile surface. Constraints for movement and usage within the system of the garment—such as the type of joining, scale, and the geometry of single modules—govern its behavior. The modulation of the body, once wrapped up, is revealed by emphasizing its stimulus potential regardless of its underlying anatomy. New body and spatial constructs are molded by the movement and interaction of participants within the garment. The soft geometry itself becomes a tactile interface between its users and the built surroundings. Choreographed Geometry was designed by Gabi Schillig with the support of an Akademie Schloss Solitude Fellowship in 2007.

CONTENTS
100% hand-stitched .1"
(3 mm) thick wool felt,
folded modules

APPLICATIONS
Textile installation,
wearable garment, textile
interface for human bodies,
transformative spatial
device

TYPES / SIZES
7.5 x 7.5' (2.3 x 2.3 m)

CONTACT
Gabi Schillig
Marienburger Strasse 29
Berlin, 10405
Germany
Tel: +49 (0)175-1705878
www.gabischillig.de
info@gabischillig.de

INTERLOCKING FABRIC TILE ASSEMBLY

In collaboration with Kvadrat, designers Ronan and Erwan Bouroullec have created Clouds, an innovative, interlocking fabric-tile concept for the home. Clouds can be used as an installation and be hung from a wall or ceiling. Clouds evolves as you add elements to it, producing a unique three-dimensional effect, coating architecture in a fluid yet chaotic way. Inspired by the inviting irregularity of the surface, one can construct a unique composition by aggregating tiles until a multifaceted, topographical surface begins to emerge. Self-expression lies at the heart of the appeal of Clouds.

The tiles, made of only one piece each, are attached by special rubber bands. Clouds can be used to create either a simple design or a complex decorative screen or wall. The tiles can be easily arranged and rearranged.

CONTENTS
Fabric, rubber

APPLICATIONS
Residential interiors, wall hanging, ceiling treatment, space divider

TYPES / SIZES
Fabrics: Tempo and Divina; Tempo color combinations: zesty orange/ brown, dark/ light blue, lime green/ slate blue; Divina color combinations: light/dark blue, light/dark gray, burgundy/deep purple and black/white; fabrics come in boxes with either 8 or 24 tiles (24-tile boxes are a mix of three colors)

CONTACT
Kvadrat A/S
Lundbergvej 10
Ebeltoft, 8400
Denmark
Tel: +45 8953-1866
www.kvadratclouds.com
kvadrat@kvadrat.dk

Cork Fabric

WOVEN CORK-BASED FABRIC

London-based designer Yemi Awosile has developed a new fabric from an innovative combination of repurposed cork and elastine. Cork Fabric is suited to interior design applications such as upholstery, wall coverings, and wall panels, and benefits from the high elasticity of elastane fiber as well as the acoustic and thermal insulating properties found in cork. The cork is made with the by-products from wine stopper production and is treated with a metallic finish. The fabric is available in a variety of colors and fiber combinations.

CONTENTS
Cork composite, elastane

APPLICATIONS
Upholstery, wallcovering, acoustic wall panels

TYPES / SIZES
Available in a variety of colors and custom sizes

ENVIRONMENTAL
Thermal insulating properties can help lower energy consumption when applied to walls

TESTS / EXAMINATIONS
Pending

LIMITATIONS
Not for external use

CONTACT
Yemi Awosile
66B Elgin Crescent
London, W11 2JJ
United Kingdom
www.yemiawosile.co.uk
yemi.awosile@network.rca
.ac.uk

Delight Cloth

LUMINOUS TEXTILE MADE WITH OPTICAL CLOTH

Delight Cloth is a light-emitting textile made with thousands of fiber-optic strands. With a diameter of only .01 to .02 inches (.25 to .5 millimeters), the optical fibers are woven into a large translucent tapestry that can be hung vertically or horizontally. The material may be used for wall or ceiling treatments, as well as for banner signage or clothing. Delight Cloth can be fabricated with embedded graphics or logos, and may be used to emit a wide variety of colors of light.

CONTENTS
Plastic optical fiber

APPLICATIONS
Wall hangings, ceiling treatments, signage, clothing

TYPES / SIZES
Fiber ø .01–.02" (.25–.5 mm); standard cloth width 5' (1.5 m)

ENVIRONMENTAL
The cloth produces minimal smoke in the event of fire

CONTACT
Lumen Co., Ltd.
8-14-17-1101 Nishishinjuku
Tokyo, 102-0076
Japan
Tel: +81352262855
www.lumen.jp
lumen@almond.ocn.ne.jp

DoubleFace

KNITTED STAINLESS-STEEL AND POLYESTER FABRIC

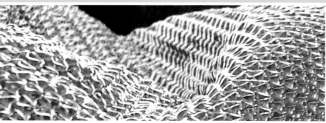 

DoubleFace is a semitransparent fabric with one side made of stainless steel and the other of polyester. This hybrid quality allows the creation of particular light effects, with reflective metal offset
by a colored background. DoubleFace is highly drapable and can be molded around various objects.
The fabric is ideal for exhibition booths, as well as residential, commercial, or institutional applications. Based on its resistance to corrosion and oxidation, DoubleFace may also be used in nautical
applications.

CONTENTS
61% AISI 316L stainless
steel, 39% polyester

APPLICATIONS
Residential, commercial,
and institutional interiors;
exhibitions

TYPES / SIZES
Maximum width 54" (140
cm); roll length 55 yards (50
m); weight .04 lbs/ft^2 (.21
kg/m^2)

LIMITATIONS
Flammable

CONTACT
Texe srl—INNTEX Div.
via Rocca Tedalda, 25
Florence, 50136
Italy
Tel: +39 055-6503766
www.inntex.com
inntex@inntex.com

Dream71

KNITTED STAINLESS-STEEL FABRIC

Dream71 is a semitransparent, ultralight fabric, developed for interior use. It is very drapable and easily conforms its shape to that of other objects. Due to the fact that it is made of 100 percent stainless steel, the textile is incombustible and may be used in public places. Dream71 is ideal for a variety of commercial and residential applications, and may also be used aboard ships due to its resistance to corrosion and oxidation.

CONTENTS
100% AISI 316L stainless steel

APPLICATIONS
Residential, commercial, and institutional interiors; exhibitions

TYPES / SIZES
Maximum width 54" (140 cm); roll length 55 yards (50 m); weight .03 lbs/ft^2 (.17 kg/m^2)

ENVIRONMENTAL
100% recyclable

TESTS / EXAMINATIONS
Dream71 fulfills the requirements of the European Class A1 for fire resistance

CONTACT
Texe srl—INNTEX Div.
via Rocca Tedalda, 25
Florence, 50136
Italy
Tel: +39 055-6503766
www.inntex.com
inntex@inntex.com

Eco Leather Tile

LEATHER FLOORING TILES

Eco Leather Tiles are made from postindustrial recycled leather and are designed for both flooring and wall applications. A new addition to the ASI Resilient collection, the material is available in 8 different colors and two textures. With the addition of rubber and acacia tree bark, Eco Leather Tiles exhibit a natural aesthetic variation that grows more attractive with use and age.

CONTENTS
Leather, rubber, acacia tree bark

APPLICATIONS
Flooring and wall applications in all commercial environments

TYPES / SIZES
12 x 12" (30.5 x 30.5 cm), 18 x 18" (45.7 x 45.7 cm); thickness 1/8" (.32 cm); 8 colors and 2 textures

ENVIRONMENTAL
Rapidly renewable construction material, recycled content, low-emission adhesives and materials

LIMITATIONS
Interior use only

CONTACT
Architectural Systems, Inc.
150 West 25th Street,
8th Floor
New York, NY 10001
Tel: 800-793-0224
www.archsystems.com
sales@archsystems.com

Freek

RUGGED CARPET

Freek is a colorful carpet that can be used for many purposes. The carpets are made of polyethylene and nylon yarns and have a pile height of 1.2 inches (3 centimeters). Good water permeability and outstanding UV stability make this carpet very durable, and thus well-suited for outdoor use. Freek has 2.250 grams fibers per square meter and a total weight of 3.865 grams per square meter. An Airlastic comfort layer supports the underside of the carpet, providing additional comfort and durability.

CONTENTS
Polyethylene and nylon
yarns

APPLICATIONS
Exterior rooms, terraces,
balconies, kitchens,
bathrooms

TYPES / SIZES
4.9 x 4.9' (1.5 x 1.5 m),
4.9 x 6.6' (1.5 x 2 m),
6.6 x 6.6' (2 x 2 m),
6.6 x 9.8' (2 x 3 m),
6.6 x 13.1' (2 x 4 m);
1.2" (3 cm) pile height;
custom sizes available

CONTACT
C&F Design BV
P.O. Box 312
Goirle, 5050 AH
The Netherlands
Tel: +31 13-530-8008
www.freek.nl
info@freek.nl

Liminal Air

INTERIOR CLOUDSCAPE

Inspired by a cave he visited at the foot of Mount Fuji, Shinji Ohmaki constructed an installation designed to render visible the air we cannot see and the sounds we cannot hear. Liminal Air is composed of 123,000 nylon strings suspended in varying lengths in such a way that they collectively create a cloudlike, inverted topography. Viewable from both the interior and exterior, Liminal Air is intended to convey to visitors the feeling of plunging into a wave of light.

CONTENTS
Nylon string, fluorescent lamp, acrylic mirror

APPLICATIONS
Art installation, ceiling treatment

TYPES / SIZES
48.5 x 8.5 x 17.7'
(14.8 x 2.6 x 5.4 m)

LIMITATIONS
Not for exterior use

CONTACT
Tokyo Gallery + BTAP
7F, 8-10-5 Ginza, Chuo-ku
Tokyo, 104-0061
Japan
Tel: +81 3-3571-1808
www.tokyo-gallery.com
info@tokyo-gallery.com

Litmuscreen

PH-SENSITIVE EXTERIOR TEXTILE SURFACE

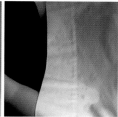

Jordan Geiger developed Litmuscreen as an architectural and product design material based on the performance and properties of litmus paper. Litmuscreen is a simple modification to an old technology—the adaptation of chemical properties found in lichen to indicate acid/base imbalances in a liquid. In this product's application, a rugged textile suitable to exterior uses displays shifts in color to red or blue as an indicator of environmental pollutants in rain. The intuitive interface provides real-time information regarding local air pollution conditions, without the use of electricity or additional construction.

CONTENTS
Lichen-based chemically imbibed, UV-resistant textile

APPLICATIONS
Awnings, canopies, tents, umbrellas

TYPES / SIZES
Custom

ENVIRONMENTAL
No energy required, allows visualization of local environmental health

CONTACT
Ga-Ga
1025 Carleton Street, Suite 14
Berkeley, CA 94710
Tel: 415-839-6926
www.ga-ga.org
info@ga-ga.org

Loop By the Yard

DO-IT-YOURSELF TEXTILE MATERIAL

Loop By the Yard is a reversible, textilelike material for all do-it-yourself projects. Designed as a free product take-back system, Loop is made from Tyvek, a spun-bonded, high-density polyethylene nonwoven material. Each yard of Loop can be cut, sewn, wrinkled, folded, pierced, hung, or hemmed like a textile. It is durable, breathable, and waterproof, making it ideal for a variety of do-it-yourself indoor and outdoor projects. The Loop By the Yard concept takes an industrial material with versatile characteristics and turns it into a product that consumers can customize. Its durability and varied applications encourage product reuse, while existing recycling infrastructure helps consumers close the recycling loop. Each order ships with a prepaid envelope for shipping scraps or the actual project (when no longer wanted or needed) back to MIO for recycling.

CONTENTS
Tyvek

APPLICATIONS
Architectural, interior design, and fashion applications; curtains, clothing, bags, window displays

TYPES / SIZES
Moire Dot pattern available in three colorways: Green-Gray, Orange-Red, and Pink-Purple

ENVIRONMENTAL
Product service system (PSS): material is sold with a prepaid envelope for shipping scraps and material back to MIO for recycling

LIMITATIONS
Not food grade

CONTACT
MIO
446 North 12th Street
Philadelphia, PA 19123
Tel: 215-925-9359
www.mioculture.com
info@mioculture.com

Morphotex

CHROMOGENIC FIBER

Morphotex is the world's first optical coloring fiber, inspired by the chromogenic principle of Morpho butterflies, which inhabit areas along the Amazon in South America. Called "living jewels," the cobalt-blue wings of Morpho butterflies impart vivid color even though they have no pigmentation.

Teijin Fibers has recreated this effect via nanotechnology, combining a total of sixty-one polyester and nylon fibers in alternating layers. By controlling the thickness of each layer, ranging from seventy to one-hundred nanometers, they can produce four basic colors (red, green, blue, and violet). Although no dyes or pigments are used for Morphotex, the process reveals a rainbow of colors according to the intensity and angle of light, due to the unique structure of the fiber. As there is no dyeing process involved in producing the fiber, Morphotex saves energy and minimizes industrial waste when compared with conventional methods.

CONTENTS
Polyester, nylon

APPLICATIONS
Textiles, clothing, upholstery, automotive, sporting goods, household appliances, portable electronics, cosmetics, optical instruments

TYPES / SIZES
Filament, short-cut fiber, and powder forms

ENVIRONMENTAL
No dyes or pigments are used, which contributes to saving energy and industrial waste

CONTACT
Teijin Fibers Limited
1-6-7, Minami-honmachi, Chuo-ku
Osaka, 541-8587
Japan
www.teijinfiber.com/english
tfj0604qa@teijin.co.jp

NanoSphere

SELF-CLEANING TEXTILE FINISH

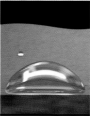

NanoSphere is a self-cleaning finish technology composed of nanoparticles that form a fine struc-ture on textile surfaces. Water or substances such as oil or ketchup simply run off the NanoSphere surface, and any residue can easily be rinsed off with a little water. Textiles with NanoSphere treat-ment require less frequent washing and can be washed at lower temperatures. They also exhibit increased abrasion and weather resistance. The function of NanoSphere is retained even after numerous washing or cleaning cycles. NanoSphere has been developed in accordance with the Bluesign standard, guaranteeing the highest possible exclusion of harmful substances.

CONTENTS

Perfluorooctane sulfonate (PFOS)- and perfluorooctane sulfonic acid (PFOA)-free C6 fluorocarbons

APPLICATIONS

Textiles, upholstery, wall coverings

ENVIRONMENTAL

PFOA- and PFOS-free, makes textiles last longer and requires less frequent cleanings, therefore saving resources.

TESTS / EXAMINATIONS

Hohenstein Research Institute testing: mechanical exposure to 5,000 abrasion cycles and 30 wash cycles; biological safety

LIMITATIONS

Lasts over 30 washings, can be "refreshed" with a heat treatment; thermally stable to 1700°C

CONTACT

Schoeller Technologies AG
Bahnhofstrasse 17
Sevelen, CH-9475
Switzerland
Tel: +41 (0)81-786-09-50
www.nano-sphere.ch
info@nano-sphere.ch

3D NONWOVEN POLYESTER TEXTILE

 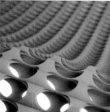  

Parametre is a lightweight, expandable three-dimensional textile system designed to deliver visual impact with minimal material. Made of 100 percent nonwoven polyester, this flexible textile can be used in a variety of residential or commercial applications in which a rich visual texture is required. Parametre may be used as a solar shading device, space divider, light diffusion panel, or screen for rapidly reconfigurable spaces.

CONTENTS
100% nonwoven polyester

APPLICATIONS
Partitions, ceilings, window treatments, dividers, screens, wall hangings/ coverings, decorative panels, light diffusion

TYPES / SIZES
Maximum panel size 14 x 16' (4.3 x 4.9 m); patterns: Mega, Hexa, Quad; colors: Terracotta, Green Apple, Cool Grey, Charcoal, Eggshell White, Canvas, Glacier Blue, Coffee Bean

ENVIRONMENTAL
100% recyclable

TESTS / EXAMINATIONS
NFPA 701-04, ASTM E 84
Class A

CONTACT
3form
2300 South 2300 West
Salt Lake City, UT 84119
Tel: 801-649-2500
www.3-form.com
info@3-form.com

MULTIDIMENSIONAL PRODUCT

Plains Collection

METAL-ENHANCED SOLAR TEXTILE

Libby Kowalski of Kova Textiles developed the Plains Collection to improve upon existing solar shade matrices. Made from vinyl-coated polyester and aluminum slit film, the Plains Collection makes use of metallic elements to improve the aesthetic qualities of this light-filtering material. The Plains Collection may be used in commercial as well as residential settings, and the fabrics may also be encapsulated in polyethylene terephthalate glycol (PETG) resin to make a rigid surface.

CONTENTS
60% vinyl on polyester, 30% polyester, 9% aluminum, 1% nylon

APPLICATIONS
Shades, panel track systems, room dividers, decorative screens, theatrical backdrops, fabrics for encapsulation in PETG resin

TYPES / SIZES
Widths: 54", 56", or 72" (137.2 cm, 142.2 cm, or 183 cm); fabric thickness: .02" (.5 mm); mesh weight: 10.7 oz/yd² (.36 kg/m²)

ENVIRONMENTAL
Polymer yarns contain recycled content

TESTS / EXAMINATIONS
NFPA 701, AATCC method 16A exceeds 60 hours, openness factor: 7.5%

LIMITATIONS
Fabrics do not easily fold or pleat in the vertical direction

CONTACT
Kova Textiles LLC
32 Union Square East, Suite 216
New York, NY 10003
Tel: 212-254-7591
www.kovatextiles.com
info@kovatextiles.com

URBAN BODY ARCHITECTURE

Public Receptors: Beneath the Skin is a textile installation by Gabi Schillig, a conceptual artist and architect in Berlin. Schillig's wearable spatial structures mediate between private users and public spaces, provoking new relationships between bodies, clothing, and the built environment. Redefining the garment as tactile architecture, Schillig developed a set of textile structures she calls "public receptors" and conducted a series of site-specific experiments for implementation in New York City. Made from felt, latex, and a variety of fastening devices, the structures are designed for attachment to specific building surfaces and street conditions, to be improvised and appropriated for clothing, furniture, habitat, or other uses.

Schillig explores the potential for soft geometries and surfaces of textiles, conventionally associated with individual bodies and human scale, to generate alternative arrangements of social space and modes of interaction in the urban fabric. For Schillig, multiple users, desires, and urban contexts are necessary to materialize her work. Designed to be interconnected and shared, her second skins evolve an architecture built upon the creativity of its participants.

Upon contact, Public Receptors transform in geometry, texture, and color from two-dimensional and often-camouflaged elements in the city to three-dimensional interfaces that sensitize and reassociate urban bodies to environments at multiple scales. *Public Receptors: Beneath the Skin* was developed with the support of a 2008–9 Van Alen Institute New York Prize Fellowship.

CONTENTS

A set of textile structures made of different types of wool felt (Karakul felt, SA Felt, BN Felt), zippers, and fastening devices (ribbons, elastic straps)

APPLICATIONS

Urban textile installation, clothing, urban furniture, shelter, fabric structures

TYPES / SIZES

3.3 x 6.6' (1 x 2 m), 4.3 x 6.6' (1.3 x 2 m); .12–.2" (3–5 mm) thick layers

CONTACT

Gabi Schillig
Marienburger Strasse 29
Berlin, 10405
Germany
Tel: +49 (0)175-1705878
www.gabischillig.de
info@gabischillig.de

MULTIDIMENSIONAL PRODUCT

Raum(Zeit)Kleider

SPATIAL TEXTILE SYSTEM

Raum(Zeit)Kleider is a spatial system that can be folded and transformed from a two-dimensional planar surface into a three-dimensional object, incorporating different functions from clothing to urban furniture and shelter. This approach seeks to produce transformations through intimate, often physical contact and participation. The use of soft materials, such as woolen felt, makes us deal with alterable materials as a membrane or second skin around our bodies, which form the walls of our own architecture and furthermore make statements about identity, desire, and intent. The garment itself becomes an interface between its users and the constructed environment, and participation becomes an important active instrument in controlling personal space. The project merges insights from architecture, fashion design, and body performance. Individuals are required to participate, eliminating both the role of the spectator and that of the author. Raum(Zeit)Kleider was supported by an Akademie Schloss Solitude Fellowship in 2008.

CONTENTS
Wool felt, zippers, push buttons

APPLICATIONS
Textile installation, body architecture, transformative object, clothing, furniture, shelter, body extension, second skin

TYPES / SIZES
ø 59.1" (150 cm); .2" (5 mm) thickness

CONTACT
Gabi Schillig
Marienburger Strasse 29
Berlin, 10405
Germany
Tel: +49 (0)175-1705878
www.gabischillig.de
info@gabischillig.de

RPET Bag

RECYCLED PET AND VINYL TEXTILES

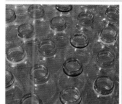   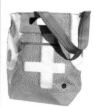

Vy&Elle creates products for everyday use from second-generation industrial waste such as reclaimed vinyl billboards, fabric banners, and 100 percent-certified recycled PET fabric. Each of their products contains 97 percent recycled postindustrial waste. Billboard vinyl is applied in its original form with little processing. The additional PET fabric is derived from plastic bottles that have been recycled into repurposed fabric, creating an alternative to non-eco fabrics and trim.

Plastic PET bottles are collected from recycling plants, washed, sorted by type and color, and then crushed. Once in a crushed state, the recycled PET plastic pellets are then heated to produce both long and short fibers for yarn production. This yarn can then be woven or knitted and used in the same way as any other fabric, thus helping to create a sustainable system for recycling waste back into everyday products such as clothing, accessories, and toys.

CONTENTS
Recycled vinyl, recycled polyethylene terephthalate (PET)

APPLICATIONS
Fashion accessories, architectural elements

TYPES / SIZES
Various

ENVIRONMENTAL
Repurposed waste materials

TESTS / EXAMINATIONS
Certification of origins and process, strength, and durability

LIMITATIONS
Stress wear, inorganic materials

CONTACT
Vy&Elle
901 North 13th Avenue, Suite 105
Tucson, AZ 85705
Tel: 520-623-9600
www.vyandelle.com
customerservice@vyandelle.com

Shutters

ELECTRONICALLY CONTROLLED TEXTILE

Shutters, designed by Marcelo Coelho, is a shape-changing, permeable textile for environmental control and communication. It is of a curtain composed of actuated louvers that can be individually addressed for precise control of ventilation, daylight incidence, and information display.

Shutters's soft mechanics are based on the electronic actuation of shape-memory alloy strands. Each strand is controlled to angle Shutters's louvers and dynamically adjust their aperture, regulating shade and ventilation, as well as displaying images and animations.

By combining smart materials, textiles, and computation, Shutters creates living environments and work spaces that are more energy efficient, while being aesthetically pleasing and considerate of inhabitants' activities.

CONTENTS
Shape-memory alloy, wool felt, conductive fibers

APPLICATIONS
Precise control of ventilation, daylight incidence, and information display

TYPES / SIZES
3 x 2' (.9 x .6 m); custom dimensions and shapes possible

ENVIRONMENTAL
Natural fibers, dynamic control of heat gain and losses

LIMITATIONS
Not for exterior use

CONTACT
MIT Media Laboratory
20 Ames Street, E15-322
Cambridge, MA 02139
Tel: 617-452-5695
www.cmarcelo.com
marcelo@media.mit.edu

Tarrot

KNITTED METAL FABRIC

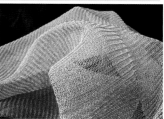

Tarrot is a knitted-fabric hybrid for interior design made of three different materials—copper, stainless steel, and brass. It has a melanged warm color with beautiful iridescent light effects. It is very drapable and can be modeled around objects. Tarrot is appropriate for exhibition booths, office and hotel decoration, and lamp shades.

CONTENTS
57% enameled copper, 22% brass, 21% AISI316L stainless steel

APPLICATIONS
Interior design

TYPES / SIZES
Maximum width 54" (140 cm); rolls of 55 yards (50 m); weight .11 lbs/ft^2 (.52 kg/m^2)

LIMITATIONS
Interior use only

CONTACT
Texe srl—INNTEX Div.
via Rocca Tedalda, 25
Florence, 50136
Italy
Tel: +39 055-6503766
www.inntex.com
inntex@inntex.com

TeXtreme

SPREAD TOW CARBON FABRIC

TeXtreme is a high-performance carbon fabric used in composites for aerospace, automotive, marine, and other industries. The distinct mechanical properties of TeXtreme and the accompanying TeXero UD tapes increase the performance and reduce the weight of composites, as well as enhance their surface finish. TeXtreme's superior performance allows for a significant weight savings in structural composite materials, making vehicles more fuel efficient.

CONTENTS
Carbon fiber

APPLICATIONS
Sports, aerospace, marine, automotive, racing, design

TYPES / SIZES
Available in a variety of fabric weights and fiber types

ENVIRONMENTAL
Lightweight, increase in fuel efficiency

CONTACT
Oxeon AB
Norrby Långgata 45
Borås, 50435
Sweden
Tel: +46 33-20-59-70
www.oxeon.se
contact@oxeon.se

Trasta

KNITTED ELECTROMAGNETIC SHIELDING FABRIC

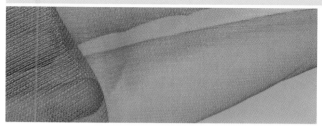

Trasta offers an effective solution for electromagnetic shielding when humidity is a fundamental factor. Trasta is very light, but at the same time is electromagnetic interference (EMI) resistant and it has very outstanding values of attenuation in a wide range of frequencies. Trasta can be used for Faraday cages, the shielding of electronic devices, and, due to its very high transparency, the shielding of video cameras. Trasta is available in fifty-five-yard (fifty-meter) rolls, and the fabric can be welded using a soldering iron.

CONTENTS
100% tin copper alloy

APPLICATIONS
EMI shielding, Faraday cages

TYPES / SIZES
Maximum width 39"
(99 cm), rolls of 55 yards
(50 m); weight .033 lbs/ft^2
(.16 kg/m^2)

ENVIRONMENTAL
Recyclable

TESTS / EXAMINATIONS
Tested according to
MIL-STD-285, it gives a
shielding efficiency of
minimum -55 dB from 100
kHz to 1GHz and minimum
-40 dB (99.9% attenuation)
from 80 MHz to 2 GHz

LIMITATIONS
Trasta is for interior use
only; maximum operating
temperature 302°F (150°C)

CONTACT
Texe srl—INNTEX Div.
via Rocca Tedalda, 25
Florence, 50136
Italy
Tel: +39 055-6503766
www.inntex.com
inntex@inntex.com

CARPET TILES WITH BUILDING-SCALED PATTERNS

A Wall Through Wall Carpet is the result of a design strategy for buildings in which the perception of space is united with the floor plan. With the use of computer-controlled carpet printing and tufting machines, it is possible to create special visual and atmospheric effects for a building in combination with a carefully executed carpeting plan.

Newly developed carpet printing techniques can be geared to replicate the repeating patterns of a design in large formats that go on endlessly. Such large repeating patterns can be fully tailored to the functional layout of a building's floor plan. In this way, every space in an office floor would occupy its own fragment of the building-scaled carpet and would therefore possess a special and unique ambience. With no one space similar to another, the experience of working in one's office or walking through corridors stimulates the sensation that the smaller spaces are part of a greater whole. This characteristic strengthens the perception of the spatial quality in the building.

CONTENTS
Wool, natural fibers, nylon, polyester, polypropylene and polyamide yarns; custom combinations available

APPLICATIONS
Flooring

TYPES / SIZES
Tiles in various sizes and qualities, ratio 1:1 or 1:2, wall-to-wall 13.1' (4 m); tufted or printed

TESTS / EXAMINATIONS
ISO standard; European EN standard; realization of carpet designs in cooperation with Enia (formerly Tarkett Somer) and architect Rau & Partners

LIMITATIONS
Minimum order quantity of 1,076 ft^2 (100 m^2)

CONTACT
Carpets for Buildings
Muyspad 15
Arnhem, 6815 EX
The Netherlands
Tel: +0031 6-27272058
www.carpetsforbuildings
.com
info@carpetsforbuildings
.com

HAND-LOOMED NATURAL GRASS-FIBER TEXTILES

Windochine's lustrous, organic weaves are hand-loomed in Asia by master craftsmen steeped in traditional techniques. Windochine's signature design is called Bamboo Silk, an ultrathin, sheer open-weave bamboo scrim. Elegant, contemporary textiles of fine natural grass fibers complement the bamboo collection. These are blends of abaca or hemp, the banana family, raffia and rattan palm fibers, wild pineapple, jute, and silk. Local plant fibers are harvested by hand, split, dried, beaten or otherwise refined, and then are knotted, spun, and handwoven into striking designs.

All plant materials are derived from renewable resources to support a sustainable environment. Most plants grow solely in the wild, and Windochine fibers are selectively harvested from mature plants to ensure quality, strength, and consistency.

CONTENTS
Abaca, raffia, jute, buntal palm, pineapple fiber, rattan skin, bamboo, silk

APPLICATIONS
Window treatments, tailored upholstery and bedding, wall coverings

TYPES / SIZES
Cut yardage, finished window treatments; stock sizes and up to 124" (315 cm) wide weaves

ENVIRONMENTAL
Renewable material resources

TESTS / EXAMINATIONS
Tested to meet NFPA 801 ratings

LIMITATIONS
Generally not for exterior use

CONTACT
Windochine
1371 El Corto Drive
Altadena, CA 91001
Tel: 626-798-8485
www.windochine.com
info@windochine.com

REPURPOSED MATERIAL

Woven Horsehair

TEXTILES FROM HORSEHAIR WEAVINGS

Marianne Kemp develops woven textiles out of unconventional combinations of materials. Using small bunches of horsehair interwoven with linen or cotton, Kemp creates curious fabrics in diverse colors and textures. The horsehair is sourced from live horses overseas, mainly from Mongolia. In the weaving process, Kemp manipulates the bunches through knotting, curling, and looping. Each final weaving expresses the unique shine, texture, and flexibility of the horsehair fiber.

Kemp explains, "I'm fascinated by the movement of the weavings, how the horsehair is manifest in the net of the weaving technique. Also, the way the light falls on the weaving plays an important role in the process. Each piece of work has its own unique character."

CONTENTS
Horsehair; cotton, silk, or wool

APPLICATIONS
Screens, wall and window panels, room dividers, lighting, installations, acoustic treatments

TYPES / SIZES
Custom

ENVIRONMENTAL
Rapidly renewable materials

LIMITATIONS
Handmade, not for exterior use

CONTACT
Marianne Kemp
Borneostraat 90 hs
Amsterdam, 1094 CP
The Netherlands
Tel: +31 (0)615254840
www.horsehairweaving
.com
info@horsehairweaving
.com

YaZa Collection

HYBRID INTERLACE SOLAR TEXTILE

   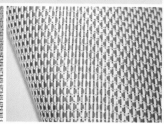

As a designer of modern polymer textiles, Libby Kowalski is dedicated to improving the aesthetic appearance of the solar shade. Her new YaZa Collection adds four new patterns to Kova's line of roller and roman shades and panel track systems. Making creative and unusual interlaces of matte, satin, and high-sheen yarns in white/silver, white/gold, and brown/copper, these textiles offer more opacity than previous Kova collections. The YaZa Collection enhances the unity of the entire interior space in hospitality, spa, and home environments and is easy to fabricate and clean.

CONTENTS
60% vinyl on polyester, 30% polyester, 9% aluminum, 1% nylon

APPLICATIONS
Roller shades, roman shades, panel track systems, room dividers, decorative screens, theatrical backdrops, fabrics for encapsulation in PETG resin

TYPES / SIZES
Width 6' (1.8 m); thickness .04" (1 mm); mesh weight 15 oz/yd² (.52 kg/m²)

ENVIRONMENTAL
Polymer yarns contain recycled content

TESTS / EXAMINATIONS
NFPA 701, AATCC method 16A exceeds 60 hours; openness factor: 8.75%

LIMITATIONS
Fabrics do not easily fold or pleat in the vertical direction

CONTACT
Kova Textiles LLC
32 Union Square East, Suite 216
New York, NY 10003
Tel: 212-254-7591
www.kovatextiles.com
info@kovatextiles.com

Zetix

BLAST-MITIGATION AND BALLISTIC-PROTECTION TEXTILES

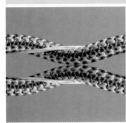

Most blast-defense systems are only capable of coping with a single explosive event. Once they have deployed, any protection they offered is lost. The unusual construction of the Zetix fabrics means that they effectively "vent" much of the blast energies through, lessening the load on the support structure. As a result, they offer a multiple-blast event solution for natural or human-initiated disasters. Zetix textiles, created by Dr. Patrick Hook, may be used for window protection, blast screens, fragmentation liners, as well as body armor enhancements.

CONTENTS
Monofilament polyester elastomers wrapped with ultrahigh molecular-weight polyethylene (UHMWPE) and woven together with ballistic nylon

APPLICATIONS
Protection against explosive events such as car bombs, IEDs, industrial accidents, hurricanes, and tornadoes

TYPES / SIZES
A variety of different sizes and thicknesses can be provided

ENVIRONMENTAL
Efficient use of materials, recyclable, improved thermal insulation

TESTS / EXAMINATIONS
In official United Kingdom government trials the Zetix fabrics have been successfully tested against simulated car bombs; other testing has shown they can also provide enhanced protection against ballistic shrapnel from such things as grenades and mortars

LIMITATIONS
Zetix is not bulletproof in the fabric state, but it can be used as an important component in hybrid armor

CONTACT
Auxetix Ltd.
P.O. Box 140, Tiverton
Devon, EX16 0AY
United Kingdom
www.auxetix.com
info@auxetix.com

09: **LIGHT**

W LIGHTING SYSTEM

INTELLIGENT PRODUCT

The 16PXL Board is a versatile, multiple-LED panel designed for interior display applications. The board unveils its full potential when placed behind semitranslucent materials or fabrics, displaying any DMX input from graphics, text, and medium-resolution video content. The high-performance LEDs spread their light in a wide-beam angle of 120 degrees across any kind of surface. The board's slim profile and light weight allow for easy integration into floors or ceilings. TX Control with advanced DMX control and Smart Chip technology enhance the 16PXL Board's performance in almost any backdrop application.

CONTENTS
LED lighting fixture

APPLICATIONS
Architectural, entertainment, retail, display lighting

TYPES / SIZES
9.1 x 9.1 x .45" (23 x 23 x 1.2 cm)

ENVIRONMENTAL
Energy efficient, with a lifetime of up to 50,000 hours and no UV radiation

TESTS / EXAMINATIONS
CE, FCC

LIMITATIONS
Indoor use only

CONTACT
Traxon Technologies
208 Wireless Centre,
3 Science Park East Avenue,
Hong Kong Science Park,
Shatin
Hong Kong, China
Tel: +852 2943-3488
www.traxon
technologies.com
marketing@traxon
technologies.com

64PXL Tile Wash

IMAGE DISPLAY LIGHTING SYSTEM

Traxon's 64PXL Tile Wash is a unique lighting element with 64 individually addressable pixels on a 19.7 x 19.7 inch (50 x 50 centimeter) surface. Its modular concept allows for large-scale installations, creating seamless surfaces of color and light as well as intricate video replays.

Controlled by the latest DMX technology, the versatile panel can display a superb array of colors, images, moving text, graphical animations, and videos. The panel's fusion effect ensures a smooth transition of lighting effects between the pixels, blending a myriad of vivid colors and patterns.

CONTENTS
LED lighting fixture

APPLICATIONS
Entertainment, retail, display

TYPES / SIZES
19.7 x 19.7 x 3 3/4" (50 x 50 x 9.5 cm)

ENVIRONMENTAL
Energy efficient with a lifetime of up to 50,000 hours and no UV radiation

TESTS / EXAMINATIONS
CE

LIMITATIONS
Indoor use only

CONTACT
Traxon Technologies
208 Wireless Centre,
3 Science Park East
Avenue, Hong Kong
Science Park, Shatin
Hong Kong, China
Tel: +852 2943-3488
www.traxon
technologies.com
marketing@traxon
technologies.com

Dune

INTERACTIVE LUMINOUS LANDSCAPE

Dune is a technocentric project applied creatively within a public space. Viewers look at, walk around, and interact with a large, undulating field of light-emitting tubes. Designed by Studio Roosegaarde, Dune is an interactive landscape that responds to the location and behavior of people. This natural/technological hybrid is represented by large numbers of fibers that are brightened according to the sound and motion of passing visitors. Studio Roosegaarde completed a recent installation of Dune that is 131 feet (40 meters) long and filled with interactive lights and sounds.

CONTENTS
LEDs, tubes, interactive electronics, software

APPLICATIONS
Interactive landscapes in architectural spaces

TYPES / SIZES
32.8' (10 m) field modules

ENVIRONMENTAL
Recyclable materials, utilizes solar power

TESTS / EXAMINATIONS
IP65, waterproof

CONTACT
Studio Roosegaarde
Nieuwe Binnenweg 256a
Rotterdam, 3021 GP
The Netherlands
Tel: 0031 182647070
www.studioroose
gaarde.net
daan@studioroose
gaarde.net

Fling

CERAMIC-TUBE LIGHT FIXTURE

Designed by Monika Piatkowski, the Fling light is made from industrial ceramic tubes used for temperature measurement in industrial furnaces. The design capitalizes on the beauty of a raw industrial material and gives it a refined aesthetic. The ring of ceramic-tube filters diffuse and direct light from a concealed central source. Clean and uncomplicated, the lamp combines functionality with an explicit sculptural simplicity.

CONTENTS
Industrial ceramic tubes,
Perspex central disc

APPLICATIONS
Lighting for domestic and
commercial applications

TYPES / SIZES
19.7 x 9.8" (50 x 25 cm);
custom sizes possible

ENVIRONMENTAL
No manufacturing involved,
reuse of industrial material

LIMITATIONS
For interior use only

CONTACT
Hive
The Studio,
6 Shardeloes Road
London, SE14 6NZ
United Kingdom
Tel: +44 (0)20-8692-0219
www.hivespace.com
hive@hivespace.com

Flip

INTERACTIVE COLOR-CHANGING LIGHT

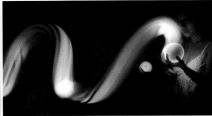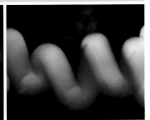

The Flip light, designed by Scott Franklin and Miao Miao of NONdesigns, is a fun and tactile new luminaire that changes color when it rotates. With no cords to tie it down, this soft rubber sphere is made for the hands as well as the desk. An equilibrium sensor invented by the firm enables every movement of the fixture to be reflected chromatically in its glow. Playing with a Flip light enables the user to find any color in the spectrum of visible light by simply rolling it, tossing it, and spinning it, creating animations of light that respond to the interactions taking place.

CONTENTS
Translucent silicone rubber, high-intensity LEDs, rechargeable battery, patented equilibrium sensor

APPLICATIONS
Interactive accent lighting

TYPES / SIZES
5" (12.7 cm) sphere

ENVIRONMENTAL
Uses no silicon chips or parts

CONTACT
NONdesigns, LLC
620 Moulton Avenue, #112
Los Angeles, CA 90031
Tel: 323-222-7883
www.nondesigns.com
info@nondesigns.com

JELLY fish

SOUND-RESPONSIVE LIGHT OBJECT

Created by Scott Franklin and Miao Miao, JELLY fish are playful objects that come to life when a viewer's presence is detected. Consisting of wire-form bodies, fan propellers, and fabric strips, the biomimetic sculptures initiate lively animations once voices, music, or other contextual sounds are detected. When the clear plastic microphone senses activity, the propellers begin to rotate, creating spinning helixes. The JELLY fish, although simple in construction, assume an autonomous state of being whose lifelike presence becomes intriguing for audiences.

CONTENTS
Microphone, sound switch, fans, fabric, light source

APPLICATIONS
Sound-responsive sculptural lighting

TYPES / SIZES
3 x 3 x 10' (.9 x .9 x 3 m) each

ENVIRONMENTAL
Lightweight construction, minimal material usage

LIMITATIONS
For indoor use only

CONTACT
NONdesigns, LLC
620 Moulton Avenue, #112
Los Angeles, CA 90031
Tel: 323-222-7883
www.nondesigns.com
info@nondesigns.com

Light Art

ECORESIN-BASED LIGHTING FIXTURES

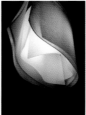

Light Art is a series of iconic light sculptures made from translucent 3form Varia Ecoresin material. Chandeliers are 3 to 4 feet (.9 to 1.2 meters) wide and 4 to 8 feet (1.2 to 2.4 meters) tall or larger, and are ideal for dramatic entry spaces or for creating a focal point over a table. Fixtures include internal fluorescent UL lighting, and cables for hanging pendants are single-bulb fixtures ranging from 14 to 40 inches (35.6 to 101.6 centimeters) in length. Multiple fixtures may be used to create a formal ensemble, and custom large-scale installations may be designed to meet the particular requirements of a space.

CONTENTS
Incandescent, fluorescent, or LED lighting; polyethylene terephthalate glycol modified (PETG), textile interlayer, aluminum

APPLICATIONS
Chandeliers, sconces, pendants

TYPES / SIZES
Custom

ENVIRONMENTAL
GREENGUARD-certified for indoor air quality, SCS certified 40% preconsumer recycled content

TESTS / EXAMINATIONS
ASTM E84: 1/4" to 3/4" gauges=CLASS B, 1" gauge=CLASS A; NFPA 286: 1/4" to 3/8" gauges=CLASS A, PASS CC1 for Light Transmitting Plastics; ANSI Z87.1 (Safety Glazing)—PASS

LIMITATIONS
Maximum temperature use 150°F (65.6°C); requires UV protection and edge sealing for exterior use

CONTACT
3form
2300 South 2300 West
Salt Lake City, UT 84119
Tel: 801-649-2500
www.3-form.com
info@3-form.com

LINEAR ELECTROLUMINESCENT LAMPS

Light Tape lamps alter one's preconceptions about point-based light sources. Imagine a lightbulb as thin as a credit card in any color that can be bent around any surface for hundreds and hundreds of feet, and that costs only a fraction of what traditional bulbs cost to operate. Developed by Electro-LuminX, Light Tape is made from Global Tungsten & Powders's light-emitting phosphors and Honeywell's encapsulant systems, and may be used for general illumination, signage, and animated displays.

CONTENTS
Honeywell barrier film, premium-quality phosphors

APPLICATIONS
Accent lighting for architectural, safety, signage, restaurant and bar, entertainment, event, and exhibit industries; backlighting for signage and various translucent materials (acrylics, onyx, and other stone)

TYPES / SIZES
Interior and exterior applications; width 1/4" (.6 cm) to 24" (61 cm); continuous lengths up to 300' (91.4 m)

ENVIRONMENTAL
Low energy consumption (a 1" x 300' piece of Light Tape consumes less power than a 100 W lightbulb and can be powered by a single power supply); no heat, gas, or UV emissions

TESTS / EXAMINATIONS
UL certified (File number E319670 project number 08CA12711), CE certified (No.0704 63147 001 and 2006/95/EC Low Voltage Directive), WEED/RoHS compliant (2002/96/EC WEEE Directive and 2002/95/EC RoHS), EMC Emissions compliant (EN55015 [CISPR15] Radiated and Conducted Emission and IEC/EN 51000-2-3 CLASS C), FR Flame Resistance compliant (EN60695-2-11:2001 in accordance with Paragraph 2. Article EC5 UTE C 12.201; Honeywell Barrier Film tested to UL746C and CSA 0.6; and UL 94VTM-0 Flame resistance), IP68 Rating for Exterior Use compliant (EN60529 + A1: 2004 and EN60598 + A1: 2003)

LIMITATIONS
Organic phosphors not currently able to emit RGB

CONTACT
Electro-LuminX Lighting Corporation
600 HP Way
Chester, VA 23836
Tel: 804-530-8536
www.lighttape.com
thelighttapeteam@
lighttape.com

Luminous SkyCeilings

BACKLIT SKY-IMAGE CEILING SYSTEM

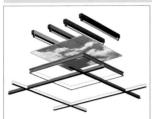

Sky Factory Luminous SkyCeilings are "authentic illusions" of real sky intended to promote comfort and vitality in viewers by transforming the experience of enclosed interior space. Hundreds of proprietary high-resolution photographs and energy-efficient, daylight-balanced lighting combine to deliver a compelling experience of nature within windowless rooms. Luminous SkyCeilings are modular and mount to existing or custom hung-ceiling grid systems. They can be configured in virtually any rectilinear configuration, as well as circular, elliptical, and custom shapes, and scaled to any size. Research attests to the value of real and simulated nature in healthcare settings as contributing to alleviation of pain, reduction in use of medications, increased patient comfort, and decreased time of hospital stays. In MRI suites, Luminous SkyCeilings have been noted to help ameliorate claustrophobic reactions. By virtue of triggering the relaxation response, Luminous SkyCeilings have been shown to increase productivity and satisfaction in the workplace.

Programmable Luminous SkyCeilings incorporate illumination designed to change over time to produce naturalistic simulations of the changing light of the day—from dawn to midday to dusk. Each of the individual lamps in the bicolor arrangement of "warm" and "cool" lights can be adjusted with a digitally addressable lighting interface (DALI), allowing for finely graduated intensity changes as well as sophisticated mixing between warm and cool lights to change the perceived color of the sky image over time.

CONTENTS
Acrylic, recycled-content aluminum, steel, T5 fluorescent lighting, pigmented inks on translucent printing substrate

APPLICATIONS
Healthcare treatment, radiology, waiting rooms, patient rooms, offices, lobbies, retail spaces, and residential spaces

TYPES / SIZES
Available in all standard ceiling-grid sizes worldwide (imperial and metric); circular, elliptical, and custom shapes available; catalog of hundreds of sky images

ENVIRONMENTAL
Energy-efficient T5 fluorescent lighting

CONTACT
The Sky Factory
801 North 18th Street
Fairfield, IA 52556
Tel: 641-472-1747 x200
www.theskyfactory.com
info@theskyfactory.com

Mitosis

INTERACTIVE LIGHTING MODULES

  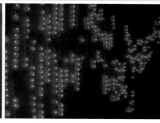

Mitosis, designed by Scott Franklin and Miao Miao, is a modular lighting system that is always changing. NONdesigns' goal was to redesign the way that lights are turned on and off and enable users to not only control the amount of light being emitted, but the patterns, shapes, and impressions that the lights create. Mitosis is intended to propel lighting from the object world into the realm of the spatial environment. Glowing colored orbs illuminate when touched once and return to their off states when touched again. Each capsule of light contains one LED and one switch that toggles it on and off when pressed. This simple and robust system invites playful interaction, bringing life to its environment.

CONTENTS
Plastic capsules containing one LED and one on/off push-button switch

APPLICATIONS
Interactive lighting installations

TYPES / SIZES
Modular chains of 2" (5.1 cm) lights

ENVIRONMENTAL
Low-energy light source

LIMITATIONS
For indoor use only

CONTACT
NONdesigns, LLC
620 Moulton Avenue, #112
Los Angeles, CA 90031
Tel: 323-222-7883
www.nondesigns.com
info@nondesigns.com

Nano Liner XB

COMPACT LINEAR LIGHTING SYSTEM

The Nano Liner XB series is a slim-profile, high-power linear fixture range equipped with 1W or K2 Luxeon LEDs. Owing to its miniaturized housing, it is ideal for space-restricting installations requiring the projection of an intense and even light output on walls or any other flat surface. The sleek design and housing of the Nano Liner XB enables discreet installation in any kind of setting. The pure lighting fixtures are free of any electronic devices within their casing, resulting in better and more reliable performance. For optimal customization, the LEDs can be connected in any desired combination: RGB, monochrome, or any other specified LED combination.

CONTENTS
LED lighting fixture

APPLICATIONS
Lighting

TYPES / SIZES
XB-9 13.6 x 1.4 x 1.5" (34.5 x 3.6 x 3.9 cm); XB-18 26.7 x 1.4 x 1.5" (67.8 x 3.6 x 3.9 cm); XB-27 39.8 x 1.4 x 1.5" (101.2 x 3.6 x 3.9 cm); XB-36 52.9 x 1.4 x 1.5" (134.5 x 3.6 x 3.9 cm)

ENVIRONMENTAL
Energy efficient lifetime of up to 50,000 hours; no UV radiation

TESTS / EXAMINATIONS
CE

CONTACT
Traxon Technologies
208 Wireless Centre,
3 Science Park East Avenue,
Hong Kong Science Park,
Shatin
Hong Kong, China
Tel: +852 2943-3488
www.traxon
technologies.com
marketing@traxon
technologies.com

3D WASHI PARCHMENT

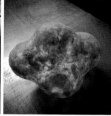

Washi architectural parchment has been made for more than five-hundred years in Japan by experienced artisan families. Washi is made from the elongated fibers of the mulberry plant, without the addition of chemicals. A few washi-producing families remain despite the forces of industrialization, and Precious Pieces has been identifying and developing contemporary applications for this particular material tradition. One such application is Orb, a three-dimentional washi parchment lampshade. Orb is especially notable for its lack of frames or supports, and its seamless spherical form belies the mark of its fabrication. One-hundred percent natural and handmade, Orb represents an intriguing modernization of a centuries-old craft.

CONTENTS
100% Japanese mulberry fibers

APPLICATIONS
Lampshades, soft sculpture, paper packages

TYPES / SIZES
Diameter from 3.5" (8.9 cm) to 31.5" (80 cm); custom shapes and patterns possible

ENVIRONMENTAL
100% natural material, 100% handmade, no chemicals used, zero emissions

LIMITATIONS
Not for exterior use

CONTACT
Precious Pieces
5 Tudor City Place, #102
New York, NY 10017
Tel: 212-682-8505
www.precious-piece.com
info@precious-piece.com

PET Wall

LUMINOUS REPURPOSED-POLYETHYLENE CURTAIN

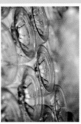

Light is essential to the realization of architecture, yet in the process of design and construction it is commonly an afterthought. Not only is the source of light important for the quality of illumination within a space, but also the materials used to capture, filter, and redirect the light.

The PET Wall is a self-supporting, luminous curtain made with thousands of postconsumer PET bottles arrayed in stacked honeycomb modules, as well as integrated-LED light nets cycling through gradually undulating sequences of warm and cool white illumination. The lightweight structure makes use of a widely disposed-of postconsumer product due to its advantageous structural and light-filtering properties. Like headlamp or light-fixture lenses, the particular thermoformed geometries of these transparent bottles convey and disperse illumination efficiently while obscuring glare. The result is a thickened surface of modular, tactile light nodes with various possibilities for programmability and interaction.

Designed by Blaine Brownell, the PET Wall is designed to expand the potential of second-use materials to the building scale. This new self-supporting "second surface" attempts to inspire a dual reading in which the viewer is simultaneously conscious of the reuse of a product as well as the ephemeral atmosphere it creates when arrayed as a large, expansive light lens.

CONTENTS
Reused Polyethylene terephthalate (PET) beverage containers, LED nets, light program control devices

APPLICATIONS
Ambient lighting, programmable wall surfaces, mood enhancer

TYPES / SIZES
Custom

ENVIRONMENTAL
Repurposed material, low-energy light source

LIMITATIONS
Self-supporting to 10' (3 m); bright light can obscure LEDs

CONTACT
Transstudio
740 Mississippi River Drive,
Suite 16D
Saint Paul, MN 55116
www.transstudio.com
info@transstudio.com

SIDE-EMITTING FIBER-OPTIC DIFFUSER SCREEN

In architecture, a doorway represents the transition from one space to another. Portal visually displays this moment of transition as a two-dimensional plane penetrated by light. As viewers watch light pass through the diffuser screen, clean lines become blurred and diffuse. Designed by James Clar, Portal celebrates the concept of a spatial threshold with abstract lines of light that travel back and forth through a door, effectively "bouncing" around a room. In addition, an integrated motion sensor detects the presence of visitors and the emitters select a random color palette for the light.

CONTENTS
Side-emitting fiber optics;
light filter diffuser screen,
fiber-optic emitter

APPLICATIONS
Lighting, interactive
installations, artwork

TYPES / SIZES
Two 150 W fiber-optic
emitter MHs; 131.2' (40 m)
of side-emitting fiber
optics; diffusion screen 6.6
x 3.3' (2 x 1 m); custom
sizes and installation
available

LIMITATIONS
Can be modified for outdoor
installation

CONTACT
Traffic Design Gallery
Saratoga Building,
Al Barsha
Dubai, 6716
United Arab Emirates
Tel: +971 4-341-8494
www.viatraffic.org
info@viatraffic.org

MULTIDIMENSIONAL PRODUCT

Softwall LED

EXPANDABLE ILLUMINATED-POLYETHYLENE PARTITION

Softwall + Softblock is a modular, space-shaping system of expandable/compressible honeycomb seating, partitions, and lighting conceived from the desire for flexible, spontaneous space-making. Opening a Softwall + Softblock element is a captivating tactile experience, as the honeycomb unfolds to create a completely freestanding structure that is hundreds of times larger than its compressed form. Softwall + Softblock forms can be resized and rearranged into almost any shape.

Softwall LED, a luminous version of Softwall + Softblock, is an integrated lighting system as flexible as its counterpart, with LED ribbons of light adjusting to changes in the wall's length and movement. Softwall LED has a soft, even glow that emphasizes the visual delicacy of the translucent textile fibers and visually enhances the expansion, contraction, and fluid movement of these completely flexible, freestanding partitions. Softwall + Softblock elements of various heights, materials, and colors all connect to one another simply and seamlessly with concealed magnets to create continuous lengths.

CONTENTS
Polyethylene nonwoven textile body, thin polypropylene end panel with rare earth magnets concealed within the last layers of the polyethylene textile body, flexible LED ribbon

APPLICATIONS
Modular system for shaping and partitioning interior space while also providing an ambient source of light, reusable system for creating temporary exhibits or private space, dampening acoustic reflections, lighting

TYPES / SIZES
Flexible in length, opens to a maximum length of 15' (4.5 m); ranges in height from 1–10' (.3–3 m)

ENVIRONMENTAL
Efficient use of material, creation of multifunctional spaces, recycled and recyclable materials

TESTS / EXAMINATIONS
NFPA 701, Class A fire rating under ASTM E84-05

LIMITATIONS
Interior use only

CONTACT
Molo Design
1470 Venables Street
Vancouver, BC V5L 2G7
Canada
Tel: 604-696-2501
www.molodesign.com
info@molodesign.com

A LANTERN MADE OF SOLAR CELLS

INTELLIGENT PRODUCT

The Solar Lampion is influenced by traditional paper lamps and structures found in nature, such as the geometric spiralling found in pinecones. Integral to its design are thirty solar cells, which form the basis for its geometry.

Designed by Damian O'Sullivan to have a minimal structure, Solar Lampion is built up of layers, or crowns, that have been cast in an aluminum alloy. Each of these crowns holds six inclined solar cells, and every crown is horizontally displaced by 30 degrees from the one below it.

This pattern results in an organic cylindrical shape, allowing the lantern to be placed in any orientation and still catch the sun's rays. The solar cells are coupled to LEDs, and these are fed by a rechargeable battery. A simple handle on top allows the Solar Lampion to be moved easily from the garden into the home, or suspended from a tree.

CONTENTS
Aluminum-alloy castings, solar cells, LEDs, electric circuit

APPLICATIONS
Residential lighting, indoor and outdoor

TYPES / SIZES
ø 7.3 (18.5 cm), height 11.8" (30 cm)

ENVIRONMENTAL
Solar-powered

CONTACT
Damian O'Sullivan Design
Burgemeester Meineszlaan
109A
Rotterdam, ZH 3022 BE
The Netherlands
Tel: +31 (0)10-425-8494
www.damianosullivan.com
info@damianosullivan.com

Sugarcube

PRISMATIC FILM LIGHTING

The Sugarcube lamp, designed by James Clar, uses a special prismatic film that separates white light into the visible spectrum of colors. A linear fluorescent-tube lamp is fitted to run diagonally from one corner to another. Because the distance between the light source and the prismatic film changes inside the cube, the Sugarcube creates the effect of colors separating and then recombining at the source.

CONTENTS
Clear acrylic cube, 12W fluorescent tube, prismatic film

APPLICATIONS
Floor and desk lamp, landscape lighting, interior lighting

TYPES / SIZES
13 x 13 x 13" (33 x 33 x 33 cm)

ENVIRONMENTAL
Long-life fluorescent-tube light, low-wattage

LIMITATIONS
Can be modified for outdoor installation

CONTACT
Traffic Design Gallery
Saratoga Building,
Al Barsha
Dubai, 6716
United Arab Emirates
Tel: +971 4-341-8494
www.viatraffic.org
info@viatraffic.org

Tensile Series

TENSEGRITY-BASED SCULPTURAL LIGHT

The works in the Tensile Series are spatial light sculptures that use tension wires to create a self-supporting structure. The wires that create tension—thus holding up the structure—also provide the electricity. The Tensile Series installations, created by James Clar, are minimal sculptures and floor lamps represented by layered, interlocking lines of illuminated color.

The wiring is connected to the fitting and not to the tube itself, allowing the user to replace the tube when it burns out without losing the structure. The user can also change color filters on the tubes, thus modifying the overall look and pattern.

CONTENTS
Fluorescent tube lamps

APPLICATIONS
Floor lighting, sculptural lighting, installation art

TYPES / SIZES
Custom wiring configuration, sizes variable

ENVIRONMENTAL
Low-wattage fluorescent tubes

CONTACT
Traffic Design Gallery
Saratoga Building,
Al Barsha
Dubai, 6716
United Arab Emirates
Tel: +971 4-341-8494
www.viatraffic.org
info@viatraffic.org

Wall Washer XB

WALL-WASHING LIGHTING SYSTEM

Wall Washer XB-18 and XB-36 are high-power LED fixtures that generate high-brightness single-color or color-changing light for a rich, wall-washing effect. The Wall Washer XB series is rated for both indoor and outdoor use, and is equipped with an extruded natural anodized-aluminum finish, which acts as an efficient heat sink for 1W and K2 Luxeon LEDs. With a wide range of available optics and beam angles, the densely packed high-brightness LEDs can be used for tightly focused and concentrated light output as well as for a long and wide light spread over increased distances.

A heat sensor within the LED casing provides additional safety via a thermal management system, which will automatically reduce the light output should the optimal working temperature be exceeded. For optimal customization, the LEDs can be connected in any desired sequence: RGB, monochrome, or any other specified LED combination.

CONTENTS
LED lighting fixture

APPLICATIONS
Lighting for architectural, retail, display

TYPES / SIZES
14.7 x 4.8 x 5.8"
(37.4 x 12.3 x 14.7 cm)

ENVIRONMENTAL
Energy efficient: lifetime of up to 50,000 hours; no UV radiation

TESTS / EXAMINATIONS
CE

CONTACT
Traxon Technologies
208 Wireless Centre,
3 Science Park East Avenue,
Hong Kong Science Park,
Shatin
Hong Kong, China
Tel: +852 2943-3488
www.traxon
technologies.com
marketing@traxon
technologies.com

WAVE MODULATION–BASED SCULPTURAL LIGHT

The sine wave, square wave, triangle wave, and sawtooth are various forms of modulations that are combined to create music, sound, electronics, and analog information. In Waveform, created by James Clar, they are represented visually, each one in a different color of neon, overlapping and combining optically as they reflect inside the box.

CONTENTS
Neon lighting, reflective glass, mirror, wooden frame

APPLICATIONS
Floor lighting, sculptural lighting, installation art

TYPES / SIZES
6.1 x 3.3 x 1' (1.85 x 1 x .3 m); 4-color neon lighting

CONTACT
Traffic Design Gallery
Saratoga Building,
Al Barsha
Dubai, 6716
United Arab Emirates
Tel: +971 4-341-8494
www.viatraffic.org
info@viatraffic.org

MULTIDIMENSIONAL PRODUCT

10: **DIGITAL**

3D PatternPrint

DIGITAL PRINTING ON MOLDABLE PLASTICS

3D PatternPrint is a new way of sublimation-printing on plastic. Designer Mary Crisp created 3D PatternPrint to bring together two-dimensional-printed planes and three-dimensional-formed shapes through engineered prints. The process chemically bonds images to plastic, allowing the plastic to be heat-formed into various shapes, with the image responding to the shape change. This is a departure from most hard-surface printing processes, as the image is applied before the final product is formed. 3D PatternPrint has been used to create a range of orb-shaped lighting objects, but could also be used for many other decorative applications.

CONTENTS
Thermoplastic, digital print

APPLICATIONS
Lighting, furniture, decorative applications

TYPES / SIZES
Varying

ENVIRONMENTAL
100% recyclable

LIMITATIONS
Angles and stretch of image have dimensional limitations

CONTACT
Mary Crisp
1 Beechbrook Avenue
Yateley, Hampshire
GU46 6LE
United Kingdom
www.marycrisp.co.uk
mary@marycrisp.co.uk

THREE-DIMENSIONAL PRINTED, CORRUGATED WOODEN MATERIAL

Wellboard is a lightweight, cellulose-based material pressed into a variety of profiles for use in exhibition, retail, and furniture design applications. 3D Printed Wellboard employs an image-transfer process developed by Okalux, called Okacolor. This innovative printing method allows digital images to maintain their geometrical integrity despite the deformations inherent in the contouring process. The combination of custom digital imagery and specific profiles can generate layered visual effects.

CONTENTS
Cellulose, digital-print
Okacolor

APPLICATIONS
Furniture production, retail,
exhibitions, interior design

TYPES / SIZES
Corrugated profile, gamma
with trapeze-shaped profile

ENVIRONMENTAL
100% recyclable

LIMITATIONS
Not for exterior use; avoid
a humid environment

CONTACT
Well Ausstellungssystem
GmbH
Schwarzer Bär 2
Hannover, D-30449
Germany
Tel: 0049 511-92881-10
www.well.de
info@well.de

INTERFACIAL PROCESS

...003

...ECT-RECOGNITION SURFACE

Digital

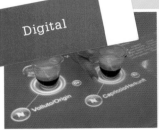

INTERFACIAL PRODUCT

The atracBar transforms the conventional counter, bar top, or information desk into an illuminated, interactive surface that will react to the objects laid on it. The multitouch surface offers a large variety of tools and possibilities. For each presented item, it is possible to insert multimedia content like comments and audio and video files into the bar's database. In retail or exhibit situations, customers or salespeople may then access all necessary product information with natural finger movements.

CONTENTS
Video-based movement tracking system, computer, beamer, screen

APPLICATIONS
Interactive display; retail, entertainment, commercial, and exhibition applications

TYPES / SIZES
Projection size 50.4 x 20.5" (128 x 52 cm); bar height 45.3" (115 cm)

ENVIRONMENTAL
Reduction in physical signage and menu materials

LIMITATIONS
Sensitive to certain lighting conditions

CONTACT
Atracsys LLC
Ch. du Ru 12
Bottens, 1041
Switzerland
Tel: +41 (0)2-533-03-50
www.atracsys.com
info@atracsys.com

INTERACTIVE WINDOW

The beMerlin interactive screen and shopping window allows unprecedented user interactivity. Thanks to a unique movement tracking system, beMerlin works with a wide range of window-glass types, even special glass used in jewelry stores or banks. The beMerlin standard configuration is composed of a video-based movement tracking system, a computer, a beamer, and a transparent screen. The cameras of the tracking system detect hand movements at a distance and send their location to the computer. The data is instantly processed and the beamer receives the information to be projected on the screen. Possible applications include urban windows transformed into interactive public-information screens, as well as virtual showrooms that allow customers to preorder items before a store opens. The beMerlin technology is invisible and easy to install, move, or transport. No expensive retrofitting or installation is required.

CONTENTS
Video-based movement tracking system, computer, beamer, transparent screen

APPLICATIONS
Public information display, virtual showrooms, interactive exhibitions, welcome windows, gaming, and entertainment

TYPES / SIZES
Monitor range from 30" (60 x 45 cm) to 100" (200 x 150 cm)

ENVIRONMENTAL
Reduces need for physical signage materials

LIMITATIONS
The optimal size and technology of the screen depends on many factors; reflected sunlight can affect hand-movement cameras

CONTACT
Atracsys LLC
Ch. du Ru 12
Bottens, 1041
Switzerland
Tel: +41 (0)2-533-03-50
www.atracsys.com
info@atracsys.com

Bloomberg ICE

INTERACTIVE MEDIA TERMINAL

Financial data and news, by its nature, can be very dry. When given the opportunity to design a showcase space for Bloomberg, however, Klein Dytham architecture (KDa) sought to transform financial information into a tangible, even playful, experience. With the collaboration of multimedia artist and interface designer Toshio Iwai, KDa designed a very public and accessible space, ICE, opposite Tokyo Station in the new heart of the Marunouchi district.

A pure white element in the space allows clouds of information to condense, like an icicle suspended from the ceiling, where data magically forms. Ice, of course, is pure and very cool, but ICE can also be interpreted as an Interactive Communication Experience. In its resting mode, ICE expresses stock tickers in a fun and easily understandable way: if the stock is up then the stock sign swells, if it drops then the stock shrinks and drops.

When you approach ICE, infrared sensors behind the glass wall detect your presence, and you begin to interact with the data. You don't actually have to touch the glass—the sensors detect you from about 12 inches (30 centimeters) away. A menu scrolls down the screen, giving you four play options: a digital harp, a digital shadow, a digital wave, and a digital volleyball.

CONTENTS
Steel frame, LED panels, glass, sensors

APPLICATIONS
Real-time information display, interactive wall

TYPES / SIZES
16.4 x 11.5' (5 x 3.5 m); four play options

LIMITATIONS
Proximity sensors do not detect physical presence beyond 12" (30 cm) from the surface

CONTACT
Klein Dytham architecture
AD Bldg 2F, 1-15-7 Hiroo,
Shibuya-ku
Tokyo, 150-0012
Japan
www.klein-dytham.
comkda@klein-dytham.com

Cartesian Wax

MULTIPLE-VISCOSITY STRUCTURAL SKIN

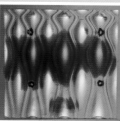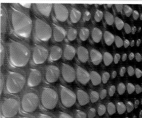

Cartesian Wax is a continuous tiling system that structurally varies across its surface area to accommodate a range of physical conditions of light transmission, heat flux, and structural support. The surface is thicker where it is structurally required to support itself, and modulates its transparency according to the light conditions of its hosting environment.

Architect and digital fabrication researcher Neri Oxman assembled twenty tiles as a continuum composed of multiple resin types—rigid and/or flexible. She designed each tile as a structural composite representing the local performance criteria as manifested in the mixtures of liquid resin.

CONTENTS

Ultralow-viscosity urethane rubber, semirigid polyurethane casting resin composite

APPLICATIONS

Advanced surface design and fabrication for product and architectural design involving structural specifications and stiffness properties

TYPES / SIZES

Custom sizes are constrained by milling bed size

ENVIRONMENTAL

Highly efficient material usage, good degree of insulation using partly natural composites

LIMITATIONS

Currently not for exterior use

CONTACT

MIT Media Laboratory
77 Massachusetts Avenue,
Building E15
Cambridge, MA 02139
www.materialecology.com
neri@mit.edu

Feather Circuit Boards

**ELECTRONIC CIRCUIT BOARDS MADE FROM CHICKEN FEATHERS
AND SOYBEANS**

Richard P. Wool has recently developed a circuit board made from soybeans and chicken feathers. A professor of chemical engineering who directs the Affordable Composites from Renewable Sources (ACRES) program at the University of Delaware, Dr. Wool seeks creative, locally available substitutes for petroleum-based resources.

"With the demise of the oil business in about twenty-five years and the ever-increasing utilization of electronic materials, it makes excellent green engineering sense to pursue new materials that are derived from renewable resources," Wool says. "The biobased materials are derived from renewable plant and animal feedstock, which use carbon dioxide from the air and help minimize global warming, as compared to petroleum feedstock."

A novel, bio-based composite material developed from soybean oils and keratin feather fibers, Feather Circuit Boards are suitable for electronic as well as automotive and aeronautical applications. Keratin fibers are a hollow, light, and tough material and are compatible with several soybean resins, such as acrylated epoxidized soybean oil (AESO). Not only is the material lighter than that of conventional circuit boards, but electrons also move at twice the speed through the feather-based printed version as well. Moreover, these materials are both bountiful in Delaware.

CONTENTS
Soybean oils, keratin
feather fibers

APPLICATIONS
Electronic, automotive, and
aeronautical applications

ENVIRONMENTAL
Bio-based petroleum
substitute, repurposed
waste product (feathers),
locally harvested materials

LIMITATIONS
The soybean resin must
be formulated to ensure
the correct dielectric loss
properties at the
frequencies used by
computers

CONTACT
University of Delaware
Center for Composite
Materials
Newark, DE 19716-3144
Tel: 302-381-3312
www.che.udel.edu/
research_groups/wool/
wool@udel.edu

RAPID-PROTOTYPED FRACTAL GROWTH TABLE

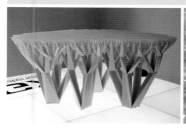

INTERFACIAL PROCESS

Rapid protoyping offers designers the possibility to explore complex forms in a direct and rapid way. CAD data is transferred directly to the prototyping machine via specific software, allowing the fabrication of most shapes.

Fractal Table is an experimental table derived from studies into fractal growth patterns. Treelike stems grow into smaller branches until they get very dense toward the top. Developers Gernot Oberfell and Jan Wertel of Platform, together with Matthias Bär, sought to create a piece that would be impossible to manufacture through another process. Fractal Table is a single piece of stereolithography apparatus (SLA) in epoxy resin, with no seams or joints.

CONTENTS
100% epoxy resin

APPLICATIONS
Protoyping, product development and testing, tooling, low-volume production or limited edition pieces

TYPES / SIZES
Maximum size 82.7 x 27.6 x 31.5" (210 x 70 x 80 cm)

ENVIRONMENTAL
No tooling required

LIMITATIONS
Very slow production, surface finish can be an issue, not all resins are UV-stable

CONTACT
.MGX by Materialise
Technologielaan 15
Leuven, 3001
Belgium
Tel: +32 16-39-61-50
www.materialise-mgx.com
info@materialise-mgx.com

Infinite-D

ANISOTROPIC TEXTURED SURFACES

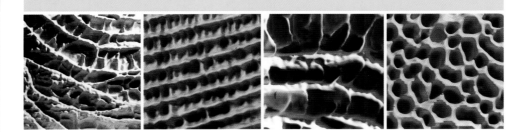

Neri Oxman visualizes, analyzes, and reconstructs two-dimensional natural microstructural tissues into three-dimensional macroscale prototypes by computing hypothetical physical responses. She uses a computational application to determine material behavior according to assigned properties and performance such as stress, strain, heat flow, stored energy, and deformation due to applied loads and temperature differences. The interaction between the directional morphology of the specimen and the tensor direction produces physical effects that emphasize the tissue's spatial texture in different ways. She then reconstructs the tissue using a CNC mill and multiple types of wood. Anisotropic in nature, grain directionality and layering are informed by the analysis resulting in laminated structural composites, which respond to given ranges of energy and loading conditions.

CONTENTS
CNC-milled laminated oak
and cherrywood

APPLICATIONS
Advanced surface design
and fabrication for product
and architectural design
involving structural
specifications and stiffness
properties

TYPES / SIZES
Custom sizes are
constrained by a net milling
bed size of 19.3 x 15.4 x 7.9"
(49 x 39 x 20 cm)

ENVIRONMENTAL
Natural composite usage
for effective stiffness

LIMITATIONS
Currently not for exterior
use, custom sizes are
constrained by milling
bed size

CONTACT
MIT Media Laboratory
77 Massachusetts Avenue,
Building E15
Cambridge, MA 02139
www.materialecology.com
neri@mit.edu

RESPONSIVE ILLUMINATED TABLE

The Interactive LED Table, designed by Because We Can's Jeffrey McGrew and Jillian Northrup, utilizes a network of twenty-four active and passive near-infrared optical sensors. The analog circuit design activates 480 superbright white LEDs to acknowledge the presence of users and objects that approach or touch the table. The Interactive LED Table is made of FSC-certified high-quality, zero-formaldehyde maple plywood and finished with a durable ultralow-VOC, water-based lacquer. There is an easily accessible on/off switch hidden on the underside of the table, and a cord that plugs into any standard 110-volt household outlet.

CONTENTS
Interactive LED panels, FSC-certified maple plywood

APPLICATIONS
Residential, retail, exhibition, entertainment

TYPES / SIZES
52 x 26 x 18" (132.1 x 66 x 45.7 cm), 40 x 26 x 18" (101.6 x 66 x 45.7 cm), 28 x 26 x 18" (71.1 x 66 x 45.7 cm), 16 x 26 x 18" (40.6 x 66 x 45.7 cm)

ENVIRONMENTAL
Low energy consumption, FSC-certified wood, low-VOC lacquer

CONTACT
Because We Can, LLC
1722 15th Street, Suite C
Oakland, CA 94607
Tel: 510-922-8846
www.becausewecan.org
us@becausewecan.org

TRANSFORMATIONAL PRODUCT

Laser-Sintered Textiles

RAPID-MANUFACTURED TEXTILES

Laser-Sintered Textiles, based on a concept by designer Jiri Evenhuis, have opened a new frontier of possibilities for the production of future textiles. Instead of creating textiles by the meter, then cutting and sewing them together into final products, Laser-Sintered Textiles could one day make needle and thread obsolete.

After several years of research in software, materials, and surface quality, Freedom of Creation (FOC) launched its first commercial products for the public in 2005. Since there were no machines made specifically for manufacturing interlocking textile patterns by layers, FOC employed rapid-manufacturing techniques such as laser sintering for their manufacture. FOC's textile products don't require any assembly and products may be made inside their own packaging.

CONTENTS
Epoxy, plaster, starch, bronze, polyamide, alumide, or steel

APPLICATIONS
Ceiling and wall hangings, upholstery, curtains, clothing, accessories

TYPES / SIZES
Maximum sizes 19.7 x 19.7 x 27.6" (50 x 50 x 70 cm) or 27.6 x 15 x 22.8" (70 x 38 x 58 cm)

LIMITATIONS
Links must be at least 1 mm thick

CONTACT
Freedom of Creation
Cruquiuskade 85
Amsterdam, 1018 AM
The Netherlands
Tel: +31 (0)20-675-84-15
www.freedomof
creation.com
info@freedomof
creation.com

Liquid Display

AMBIENT LIQUID-BASED SCREEN

Digital

Liquid Display is an ambient display composed of a screen filled with liquid and programmed air bubbles. Developed by Nicolas Büchi, the interface consists of multiple cubes filled with water or water-based emulsion. The Liquid Display is intended to convey information discreetly—approaching the limits of perception—as opposed to the increasingly saturated information exhibited by contemporary display technologies.

CONTENTS
Water or water-based emulsions, electricity, air

APPLICATIONS
Ambient display, visual interface

TYPES / SIZES
27.6 x 39.4 x 2" (70 x 100 x 5 cm); custom sizes possible

TESTS / EXAMINATIONS
Bachelor-degree project, juried with the quality seal of the Europrix Multimedia Awards.

CONTACT
Nicolas Büchi
Wettsteinplatz 4
Basel, BS 4058
Switzerland
Tel: +41 (0)78-665-51-44
www.theliquiddisplay.com
buechi@winterlife.com

RESPONSIVE BUILDING-INFORMATION NETWORK

Living City is a platform for buildings to talk to one another. It involves small modules that attach to existing building facades, collect sensor data, transmit it to other buildings, compare local and remote information, and change facade porosity or display as a result. Living City allows buildings to share information about environmental data such as air quality. The system monitors local indoor and outdoor contaminant levels, compares them to levels at nearby and remote buildings, and opens or closes gills in the facade to adjust airflow and offer a visual display of environmental conditions. The platform is open-ended, and other sensors or actuators can be swapped in, allowing buildings to exchange a wealth of information that they are already collecting.

Developed by David Benjamin and Soo-in Yang, the Living City system represents an important step in the evolution of responsive buildings. Currently, even the most sophisticated buildings only connect local input with local output. Using Living City, buildings can expand to connect local and remote input with local and remote output. The building facade becomes a location of data sensing, communication, and responsive performance and display—imbuing the city with a new layer of interactivity.

CONTENTS
Wireless sensor network, carbon monoxide sensor, nitrogen dioxide sensor, pins for integration of other sensors, custom circuit-board, rechargeable battery, waterproof casing, custom software, internet connection, custom control board for multiple outputs

APPLICATIONS
Deployment on facades of residential, commercial, or institutional buildings; input can be air quality, energy usage, or other

sensor data; output can be the dynamic opening of a facade to allow airflow, scheduling of energy consumption, or other changes to local conditions

TYPES / SIZES
Sensor nodes 4.5 x 4.5 x 1" (11.4 x 11.4 x 2.5 cm) or 8 x 4 x 3" (20.3 x 10.2 x 7.6 cm); custom sizes available

ENVIRONMENTAL
Monitors air quality, raises awareness of local and global environmental conditions, can be adapted to monitor energy usage

in multiple buildings and reduce overall energy consumption

LIMITATIONS
System currently requires customized setup and consultation; batteries for sensor nodes require recharging

CONTACT
The Living
146 West 29th Street, #4RE
New York, NY 10014
www.thelivingnewyork
.com
life@thelivingnewyork.com

Media Cubes

TANGIBLE MEDIA CONTROLS

Media Cubes invite tangible interaction to control media. Unlike conventional remote devices, this system consists of two cubes made out of wood. The system can track the orientation and rotation of the objects in order to relay particular commands. The user controls a particular function by orienting its symbol up, then rotating the cube to adjust the function—such as changing the volume level or scrolling through a song list.

The technology inside the cube is fairly simple. Gyroscopic and accelerometric sensors detect movement with a high level of accuracy. The signal is then sent via radio technology, the same way a wireless mouse works. The cubes are charged with contactless induction technology, and there is no need for a power input. In this way, the Media Cubes demonstrate the unexpected integration of technological controls with a more intuitive, tactile interface.

CONTENTS
Wood, gyroscopic and accelerometric sensors, electronics

APPLICATIONS
Remote media control

TYPES / SIZES
1.8 x 1.8 x 1.8"
(4.5 x 4.5 x 4.5 cm)

ENVIRONMENTAL
No power requirement

CONTACT
Mattias Andersson
Palandergatan 17
Johanneshov, 12137
Sweden
Tel: +46 70-6716012
www.mattiasandersson
.com
info@mattiasandersson
.com

Monocoque

SINGLE-SHELL PHOTOPOLYMER SKIN

Monocoque (the French word meaning "single shell") is named for a construction technique that supports structural load using an object's external skin. Contradictory to the traditional design of building skins, which distinguishes between internal structural frameworks and nonbearing skin elements, this approach promotes the heterogeneity of material properties.

Monocoque, developed by Neri Oxman, has a structural skin that is generated using a Voronoi pattern, the density of which corresponds to simulated loading conditions. The distribution of shear-stress lines and surface pressure is embodied in the allocation and relative thickness of veinlike elements built into the skin. The prototype model was three-dimensionally printed using Objet's PolyJet Matrix technology, which allows for the assignment of structural properties to multiple three-dimensional printed substances. This innovative technology allows parts and assemblies made of multiple materials to be printed within a single build, as well as creating composite materials that present preset combinations of mechanical properties.

CONTENTS
100% composite photopolymers

APPLICATIONS
Advanced surface design and fabrication for product and architectural design involving structural specifications

TYPES / SIZES
Custom sizes are constrained by a net bed size of 19.3 x 15.4 x 7.9" (49 x 39 x 20 cm)

ENVIRONMENTAL
Highly efficient material usage, may be used for light control

LIMITATIONS
Currently not for exterior use, size is limited to net tray dimensions

CONTACT
MIT Media Laboratory
77 Massachusetts Avenue,
Building E15
Cambridge, MA 02139
www.materialecology.com
neri@mit.edu

SOUND-RESPONSIVE WOOD PLANKS

INTERFACIAL PRODUCT

PLANKS are the result of collaboration between artist Henrik Menné and scientist Anna Vallgårda as an exploration into the aesthetic potential of computational composites. Composed of wood boards and electronic circuitry on a supporting frame, PLANKS responds to local stimuli as an experiment in animating architectural enclosures.

When the sonic activity near a PLANK rises above a certain threshold, the board will gradually bend outward; as the room grows quiet again, the PLANK will gradually straighten out. In this way, PLANKS collectively will cause a room to "shrink" as the sonic activity rises and "expand" as it falls. Each PLANK works individually with adjustable sensitivity and expression, and the combination of multiple PLANKS conveys an emergent, lifelike sensibility.

CONTENTS
Pine planks, microphones, servomotors, Arduino circuit boards, supporting structure

APPLICATIONS
Environmentally sensitive interior surfaces or installations

TYPES / SIZES
Each plank 3.9 x 78.7"
(10 x 200 cm)

LIMITATIONS
Not for exterior use; requires a 12V power source; prototype only

CONTACT
Anna Vallgårda
Rued Langgaards Vej 7
Copenhagen, 2300
Denmark
www.akav.dk
akav@itu.dk

Raycounting

LIGHT-MAPPED STRUCTURAL SURFACES

  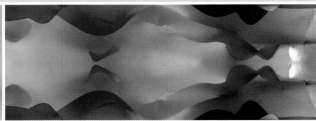

Raycounting is a method for generating form by registering the intensity and orientation of light rays. Three-dimensional surfaces of double curvature are the result of assigning light parameters to flat planes. Developed by Neri Oxman, the algorithm calculates the intensity, position, and direction of one or multiple light sources placed in a given environment, and assigns local curvature values to each point in space corresponding to the reference plane and the light dimension. The models explore the relationship between geometry and light performance from a computational geometry perspective. Light performance analysis tools are reconstructed programmatically to allow for morphological synthesis based on intensity, frequency, and polarization of light parameters as defined by the user.

CONTENTS
Nylon

APPLICATIONS
Advanced surface design and fabrication for product and architectural design involving structural specifications and stiffness properties, versatile for industrial applications

TYPES / SIZES
Custom sizes are constrained by selective laser sintering (SLS) printing bed size

ENVIRONMENTAL
Highly efficient material usage, material produces unique qualities

LIMITATIONS
Currently not for exterior use, custom sizes are constrained by printing bed size

CONTACT
MIT Media Laboratory
77 Massachusetts Avenue,
Building E15
Cambridge, MA 02139
www.materialecology.com
neri@mit.edu

INTELLIGENT FAUCET

INTERFACIAL PRODUCT

The Ripple Faucet offers a new way to interact with water. Drawing inspiration from the ripples on water's surface, the faucet creates a strong visual relationship with the water flow and temperature. The faucet's visual system is composed of the ripple interface surface, two channels from which water flows, and a metal ball that sits on top of the rippled surface and is used to control the faucet.

The movement of the ball controls the flow and the temperature: moving away from the center increases flow, and moving around the circle controls the temperature. LEDs under the ripple surface light up with varying intensity and color according to the flow and water temperature to provide the user with additional visual feedback. Under the ripple surface, there is an array of sensors and electromagnets that provide tactile feedback to the user as they move the ball to discrete points on the surface.

The Ripple Faucet is a result of collaboration between Smith Newnam, an industrial designer based in North Carolina, and Shimon Shmueli, a partner with Touch360, a product design and innovation consultancy.

CONTENTS
Polycarbonate, electromagnetic sensor nodes, electronic flash heater, cast aluminum, servo-controlled valve, RGB LEDs

APPLICATIONS
Residential sinks

TYPES / SIZES
Fits custom sinks

ENVIRONMENTAL
Minimization of water waste via a flash heater

CONTACT
Smith Newnam
622 Porter's Neck Road
Wilmington, NC 28411
Tel: 919-225-1318
www.smithnewnamdesign.com
smith@touch360.com

Sensetable

SENSING SURFACE FOR INTERACTIVE APPLICATIONS

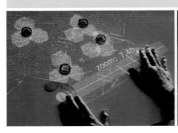  

Sensetable is a hardware platform for creating an interactive digital experience on a tabletop surface. It senses the movements of objects on its surface and uses these movements to control a computer. A variety of interactive applications have been developed for the Sensetable such as Audiopad, a new type of music DJ system, and the Reactions Table, a science museum exhibit about the periodic table of the elements. Objects on the tabletop become the interface to custom computer software. For example, in the Audiopad system, each object on the table represents a piece of sound. One moves these objects on the table to control the various parameters of musical synthesis.

Patten Studio offers custom software development for the Sensetable, and an open-source software development kit is available. A standard Sensetable antenna panel measures 9.5 x 14.6 inches (24 x 37 centimeters), and these panels can be tiled to create an interactive surface of the desired size. The antenna panels can be built into a variety of surfaces for permanent use, or placed on top of any ordinary tabletop.

CONTENTS
Electronic circuitry

APPLICATIONS
Interactive kiosks, tables, and walls

TYPES / SIZES
Modular panel 9.5 x 14.6"
(24 x 37 cm); custom shapes and sizes available

ENVIRONMENTAL
Designed for easy disassembly, reprogramming, and reuse; software design based on open standards to postpone hardware obsolescence

TESTS / EXAMINATIONS
RoHS-compliant, lead-free

LIMITATIONS
Cannot be installed underneath metal surfaces

CONTACT
Patten Studio LLC
33 Flatbush Avenue
Brooklyn, NY 11217
Tel: 617-229-5736
www.pattenstudio.com
info@pattenstudio.com

SLAP Widgets

TRANSPARENT TANGIBLE CONTROLS FOR MULTITOUCH TABLES

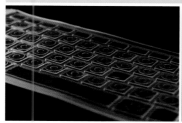

SLAP Widgets are tangible controls such as knobs, sliders, buttons, and keyboards for interactive tables. They are made of transparent materials like acrylic or silicone. Placed onto a multitouch table, the table recognizes the SLAP Widgets via markers and displays the appropriate information beneath them, such as a graphical keyboard layout underneath the physically transparent SLAP Keyboard. SLAP Widgets combine haptic feedback of physical objects with the flexibility of graphical user interfaces, such as dynamic changing labels. For example, when pressing the command key on the physical SLAP Keyboard, the table displays all invisible shortcut keys, such as copy and paste.

SLAP Widgets are general purpose tools and support manipulating graphical objects on the table. The visual appearance of the widgets depends on the graphical object they are connected with. These connections can be established and changed on the fly. A SLAP Knob connected to a movie, for example, visually changes to a navigation knob to step frames. Connected to an image, its visual appearance changes to a menu to manipulate image parameters. With SLAP Widgets, people feel the form and structure of media interaction, which improves input accuracy. Simultaneously, the transparent physical controls reveal different manipulation possibilities.

CONTENTS
Mixed-cast acrylic and silicone, or entirely cast from silicone (SLAP Keyboard)

APPLICATIONS
Providing haptic feedback on planar multitouch surfaces, enhancing typing performance on planar virtual keyboards, providing eyes-free and both-handed interaction on tabletops

TYPES / SIZES
SLAP Keyboard 11 x 4.5" (28 x 11.5 cm), SLAP Knob 3.9 x 3.9" (10 x 10 cm), SLAP Keypads 3.2 x 2" (9 x 5 cm) or 2.2 x 2" (5.5 x 5 cm), SLAP Slider 5.5 x 1.6" (14 x 4 cm)

ENVIRONMENTAL
No need for any electronics or batteries

LIMITATIONS
Requires a vision-based multitouch table

CONTACT
RWTH Aachen University, Media Computing Group Ahornstrasse 55 Aachen, Nordrhein-Westfalen 52074 Germany Tel: +49 241-80-21051 hci.rwth-aachen.de/slap slap@cs.rwth-aachen.de

Unrepeatable Carpets

CARPET TILES WITH NONREPEATING PATTERNS

    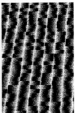

A collaboration between visual artist Marcel Kronenburg and software engineer Marten Teitsma, Unrepeatable Carpets are the result of a process designed to create unique carpet patterns throughout buildings. A reaction against the repetitive monotony of standard carpet tiles, Unrepeatable Carpets apply randomly generated images and patterns to a variety of carpet materials. Custom-designed software runs a computer-controlled carpet printing machine, and this process generates an endless variety of outcomes using a particular decorative pattern. Due to the universal quality of the pattern, however, tiles may still be easily replaced when necessary.

CONTENTS
Wool, natural fibers, nylon, polyester, polypropylene and polyamide yarns; custom combinations available

APPLICATIONS
Flooring

TYPES / SIZES
Tiles in various sizes, ratio 1:1 or 1:2, wall-to-wall 13.1' (4 m); tufted or printed

TESTS / EXAMINATIONS
ISO standard; European EN standard; realization of carpet designs in cooperation with Brintons, United Kingdom, tested by Zimmer Machinebau, Austria

LIMITATIONS
Minimum order quantity of 100 m²

CONTACT
Carpets for Buildings
Muyspad 15
Arnhem, 6815 EX
The Netherlands
Tel: +0031 6-27272058
www.carpetsforbuildings
.com
info@carpetsforbuildings
.com

Vertebrae

DIGITALLY FABRICATED CORIAN BENCH

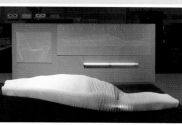
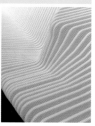

Vertebrae is a kit of parts that allows the construction of multiscale objects that modify the spaces they inhabit. This piece, while sculptural in form, proposes areas to rest and sit while defining separate spaces within a room. The segments of Vertebrae are constructed of DuPont Corian Illumination Series, with embedded lights, creating a soft diffuse glow that appears to emanate from within.

Designed by Douglas Birkenshaw of B+H Architects, the form of Vertebrae was derived using three-dimensional digital tools in an iterative process that allowed successive studies of various forms. The primary tool was a lofting program that generated forms from just a few "stations." These stations became ergonomic target points to create the desired functionality and aesthetic quality in an otherwise amorphous form. The end shapes were generated by taking snapshots of an animation of the shapes morphing along a trajectory. These resulting segments were fed directly into the software for three-axis milling machine fabrication. Each shape was attached to the next using a flexible joint creating the curvaceous, segmented form.

CONTENTS
DuPont Corian, stainless-steel threaded rod, T5 fluorescent lights

APPLICATIONS
Indoor or outdoor seating

TYPES / SIZES
180 x 30 x 30" (457 x 76.2 x 76.2 cm); custom variations possible; thickness 1/2" (1.3 cm)

CONTACT
B+H Architects
481 University Avenue,
Suite 300
Toronto, ON M6R 1V6
Canada
Tel: 416-596-2299
www.bharchitects.com
email@bharchitects.com

TRANSFORMATIONAL PRODUCT

Wind 3.0

INTERACTIVE VENTILATOR WALL

Wind 3.0 is an interactive wall feature composed of hundreds of fibers that respond to a viewer's presence based on a connection between electronic sensors and ventilators. Wind 3.0 moves with the viewer—when there is a lot of activity the wall makes large fluid motions, while in other circumstances the fiber animation resembles a soft breeze. In this way, a direct relationship is made between human behavior and sculptural dynamics. Developed by Netherlands-based Studio Roosegaarde, Wind 3.0 plays with the similarities and differences between nature and technology.

CONTENTS
Ventilators, tubes,
interactive electronics
and software

APPLICATIONS
Interactive surfaces, art
installation

TYPES / SIZES
13.1 x 1.6 x 7.7'
(4 x .5 x 2.35 m)

ENVIRONMENTAL
Recyclable materials

TESTS / EXAMINATIONS
IP65

CONTACT
Studio Roosegaarde
Nieuwe Binnenweg 256a
Rotterdam, 3021 GP
The Netherlands
Tel: 0031 182647070
www.studioroosegaarde
.net
daan@studioroosegaarde
.net

INDEX

DESIGNER INDEX

Abell, Ian, 50
Abell, Richard, 50
Andersson, Mattias, 233
Atkins, Kelly, 106
Awosile, Yemi, 78, 170

Bär, Matthias, 227
Barban, Gianfranco, 113, 134
Bardell, Freya, 116
Benjamin, David, 232
Berbet, Marie-Virginie, 117
Birkenshaw, Douglas, 241
Bouroullec, Ronan and Erwan, 133, 169
Brand, Michelle, 102
Brodarick, Gregg, 113, 134
Browka, Edward, 81, 115
Brownell, Blaine, 210
Büchi, Nicolas, 231

Clar, James, 211, 214, 215, 217
Coelho, Marcelo, 159, 186
Covelli, Fabrice, 94
Crisp, Mary, 220

Davidson, Stephanie, 36, 42
de Jong, Reinier, 83
Dhinojwala, Ali, 51
Dry, Carolyn, 23, 35, 61, 122
Dytham, Mark, 224

Etkin, Suzan, 138, 147

Franklin, Scott, 160, 202, 203, 207
Fukao, Tokihiko, 21

Geiger, Jordan, 177
Giles, Harry, 124
Giostra, Simone, 62, 145

Herzig, Thomas, 119
Hoberman, Chuck, 48, 52, 53, 54, 56, 93
Hook, Patrick, 194
Howe, Brian, 116

Iwai, Toshio, 224

Kemp, Marianne, 192
Kirk, Duncan M., 142
Kivley, Blane, 148
Klein, Astrid, 224
Kowalski, Libby, 182, 193
Kronenburg, Marcel, 240

Lath, Abhinand, 22, 24, 108, 121
Laurysen, Yvonne, 70
Lixfeld, Alexa, 18
Loebach, Paul, 84

Mantel, Erik, 70
Marques, Fernando Miguel, 151
Marquina, Nani, 96
McDonald, Ben, 40
McGrew, Jeffrey, 229
Menné, Henrik, 235
Meredith, Neil, 30, 39, 44
Miao Miao, 160, 202, 203, 207
Miquel, Ariadna, 96
Mirtsch, Frank, 64

Nelson, Ralph, 16
Newnam, Smith, 237
Nilsson, Parisima, 38
Northrup, Jillian, 229

Oberfell, Gernot, 227
Ohmaki, Shinji, 112, 176
Olsen, Eric, 33, 126
O'Sullivan, Damian, 213
Oxman, Neri, 225, 228, 234, 236

Patten, James, 238
Piatkowski, Monika, 80, 201

Rafailidis, Georg, 36, 42
Roosegaarde, Daan, 200, 242

Satterfield, Blair, 99
Schillig, Gabi, 168, 183, 184
Shmueli, Shimon, 237
Swackhamer, Marc, 99

Teitsma, Marten, 240
Tomé, Peter, 95

Vallgårda, Anna, 235
van Abbema, Jelte, 161

Wertel, Jan, 227
Wool, Richard P., 97, 226

Yang, Soo-in, 232

Zhou, Hongtao, 86

MANUFACTURER INDEX

3form, 132, 149, 150, 181, 204

Agriboard Industries, 69
Alcoa Architectural
    Products, 60
AlexaLixfeld Design
    GmbH, 18
Ambient Glow Technology, 95
Amico, 49
Architectural Systems, Inc.,
    32, 43, 73, 79, 82, 87, 104,
    110, 111, 141, 174
Atracsys LLC, 222, 223
Auxetix Ltd., 194

Based Upon, 50
Because We Can, LLC, 229
B+H Architects, 241
B.lab Italia, 113, 134
Brembo, 31
BubbleDeck, 17

Cambridge Architectural, 63
Carpet-Burns Ltd., 106
Carpets for Buildings,
    190, 240
C&F Design BV, 175
Composite Technologies
    Corporation, 25
Contrarian Metal
    Resources, 55

Damian O'Sullivan
    Design, 213
DeepRoot, 123
Duo-Gard Industries Inc., 92
DXD Inc., 135

Ecovative Design, 81, 115
Electro-LuminX Lighting
    Corporation, 205
Engineered Timber
    Resources, 85
EverGreene Architectural
    Arts, Inc., 144
Expancel/Eka Chemicals
    Inc., 101
Experimonde, 118

FormFormForm Ltd., 103
Fproduct, 94
Freedom of Creation, 230

Ga-Ga, 177
Greenmeme, 116
GreenPix LLC, 145

Henkel AG & Co. KGaA, 127
Hive, 80, 201
Hoberman Associates, 48, 52,
    53, 54, 56, 93
HOG Works Pty. Ltd., 120
HolzBuild, 71
HouMinn Practice, 99
Hycrete, Inc., 19

Imbrium Systems, 41
Innovative Glass Products,
    146
Intaglio Composites, 59

Kaynemaile Ltd., 109
Kebony ASA, 74
Kirei USA, 72, 75, 76
Klein Dytham architecture,
    224
Kova Textiles LLC, 182, 193
Kvadrat A/S, 169

La Casa Deco, 34
LAMA Concept, 70
Lamin-Art, 156
LCP Technology GmbH, 157
Living, The, 232
Loom, 16
Lumen Co., Ltd., 171

Matter Practice, 130
Meld, 26
Meshglass Ltd., 142
Metaalwarenfabriek Phoenix
    BV, 57
.MGX by Materialise, 227
MIO, 178
MIT Media Laboratory, 159,
    186, 225, 228, 234, 236
Molo Design, 212
Moving Color, 148

Nanimarquina, 96
Nathan Allan Glass Studios
    Inc., 139, 140, 143, 153
Natural Process Design Inc.,
    23, 35, 61, 122
Noguchi Laboratory,
    University of Tokyo, 21
NONdesigns, LLC, 160, 202,
    203, 207

Oxeon AB, 188

Panelite, 98
Parisima Kadh AB, 38
Patten Studio LLC, 238
Pentstar, 37
Plastipack Limited, 131
Precious Pieces, 158, 162,
    163, 209
Reholz GmbH, 68

Reynolds Polymer Technology,
    Inc., 100, 107, 114, 129
Rieder Faserbeton-Elemente
    GmbH, 20
Robin Reigi Inc., 152
Rubbersidewalks, Inc., 128
RWTH Aachen University,
    Media Computing Group,
    239

Schoeller Technologies AG,
    180
Sensitile Systems, 22, 24, 108,
    121
Sheet Design, 30, 39, 44
Shimizu Corporation, 58
Sky Factory, 206
SMIT, 105, 125
SolPix LLC, 62
Studio Roosegaarde, 200, 242
Superficial Studio, 33, 126
Suzan Etkin Enterprises,
    138, 147

Teijin Fibers Limited, 179
Texe srl—INNTEX Div., 172,
    173, 187, 189
The Greenhaus Ltd., 102
Tokyo Gallery+BTAP, 112, 176
Touchy-Feely, 36, 42
Traffic Design Gallery, 211,
    214, 215, 217
Transstudio, 210
Traxon Technologies, 198,
    199, 208, 216

University of Akron, 51
University of Delaware,
    97, 226
University of Wisconsin-
    Madison, 86

Vitra, 133
Vy&Elle, 185

Well Ausstellungssystem
    GmbH, 77, 164, 221
Windochine, 191
Wovin Wall, 88

PRODUCT INDEX

12 Blocks, 16
16PXL Board, 198
24K Blown Glass, 138
3D PatternPrint, 220
3D Printed Wellboard, 221
3D Veneer, 68
3S Solar Block, 92
64PXL Tile Wash, 199

Adaptive Fritting, 93
Adaptive Shading, 48
Agriboard, 69
Aire Pad, 94
Ambient Glow Technology, 95
APEX Mesh, 49
atracBar, 222

BAMBOO, 70
Based Upon Surfaces, 50
BBS Panels, 71
beMerlin, 223
Bicicleta, 96
Bio-Based Foams, 97
Bloomberg ICE, 224
BubbleDeck, 17

CarbonCoat, 51
Cartesian Wax, 225
Cassini Blocks, 30
Catalyst, 156
Ceramic Composite Material, 31
Choreographed Geometry, 168
ClearSeam ITL, 98
Cloak Wall, 99
Clouds, 169
Coco Tiles, 72
Convex Series—Shapes, 139
Cork Fabric, 170
Crackle, 140

Creacrete, 18
Crystallized Glass Stone, 141
Crystalmeshglass, 142

Delight Cloth, 171
DoubleFace, 172
Dream71, 173
Dune, 200
Dura, 100

Eco-Essence, 32
Eco Leather Tile, 174
Ecolinea, 73
Electroboard, 33
Element, 19
Emergent Surface, 52
Expancel, 101
Expanding Helicoid, 53

Feather Circuit Boards, 226
FibreC, 20
Flexipane, 34
Fling, 201
Flip, 202
Flowerfall, 102
Fly Ash Panels, 35
Formerol, 103
Found Space Tiles, 36
Fractal Table, 227
Freek, 175
Fusion, 143

Graphic Panels, 104
GreenPix, 145
Grow, 105
G Series, 144

Heat Treated Carpet, 106
Helicone HC, 157
Hoberman Arch, 54

Ice, 107
Infinite-D, 228
Interactive LED Table, 229
InvariBrush, 55
Iris Dome, 56

Jali Cascata, 108
JELLY fish, 203

Kaynemaile, 109
Kebony, 74
KiloLux, 146
Kirei Bamboo, 75
Kirei WheatBoard, 76
Kozo, 110
Kraftplex, 77

Laser-Cut Cork, 78
Laser-Sintered Textiles, 230
Light Art, 204
Light-Sensitive Concrete, 21
Light Tape, 205
Liminal Air, 176
Linea Nova, 79
Liquid Display, 231
Litmuscreen, 177
Living City, 232
Loop By the Yard, 178
Luminous SkyCeilings, 206
Lunalite, 111

Made to Measure, 80
Media Cubes, 233
Memorial Rebirth, 112
Mercury Glass, 147
Metal Series, 113
Mirage, 114
Mitosis, 207
Monocoque, 234
Morphotex, 179

Mycobond, 81
Mycoply, 115
MYPODLife, 116

Nano Liner XB, 208
NanoSphere, 180
Narco, 117
No-Limit Panels, 57
Northern Lights, 148

OneStep Building System, 37
Orb, 209

Parabienta, 58
Parametre, 181
PET Wall, 210
Photo-Engraved Aluminum, 59
PIXA, 22
Placage, 82
Plains Collection, 182
PLANKS, 235
Pneu-Green Facade, 118
Pneumocell, 119
Portal, 211
Poured Glass, 149
Premium Parchment, 158
Pressed Glass, 150
PS Photo Tiles, 38
Public Receptors, 183
Pulp-Based Computing, 159

Q Blocks, 39

Rainwater H$_2$OG, 120
Raum(Zeit)Kleider, 184
Raycounting, 236
Recycled Crystal Glass, 151
REK, 83
Reynobond with Kevlar, 60

Ripple Faucet, 237
RPET Bag, 185

Scintilla Lumina, 121
Self-Repairing and Sensing
    Matrices, 122
Self-Repairing Composites, 61
Self-Repairing Concrete, 23
Sensetable, 238
Sensitive Apertures, 40
SentryGlas Expressions, 152
Shelf Space, 84
Shutters, 186
Silva Cell, 123
SITumbra, 124
SLAP Widgets, 239
Softwall LED, 212
Solar Ivy, 125
Solar Lampion, 213
Solar Water Tarp, 126
SolPix, 62
Solucent, 63
Songwood, 85
SorbtiveMEDIA, 41
SOUND Wall, 160
Stax, 153
Structural Tambour, 86
Sugarcube, 214
Symbiosis, 161

Tarrot, 187
Tatami Igusa, 162
Tensile Series, 215
Terocore, 127
Terrazzo Lumina, 24
Terrewalks, 128
TeXtreme, 188
Textures, 129
Thermomass, 25

Topiary Tile, 130
Trasta, 189
Trigger Point, 42

Unrepeatable Carpets, 240

VapourGuard, 131
Varia Ecoresin, 132
Vault-Structured Metal, 64
Vegetal Chair, 133
VenCork, 87
Vertebrae, 241

Wall Through Wall Carpets, 190
Wall Washer XB, 216
Washi Laminated Glass, 163
Watercolors, 134
Wave Wall, 88
Waveform, 217
Wellboard, 164
Wind 3.0, 242
Windochine, 191
Woven Horsehair, 192
Woven Stone, 43

Xposed, 26

YaZa Collection, 193
Y Blocks, 44

Zcell, 135
Zetix, 194

ACKNOWLEDGMENTS

I would like to thank Nora Ronningen and Heather Brownell for their invaluable assistance in compiling this collection.

I am indebted to Renee Cheng, Thomas Fisher, and the rest of the faculty and staff at the University of Minnesota School of Architecture for their support of my work.

I would also like to thank Tom Buresh and Douglas Kelbaugh for supporting my research endeavors at the University of Michigan, as well as Chris Drinkwater and Natasha Krol for their assistance with the Fellows Exhibition.

I deeply appreciate the critical support given by Amy Cortese with the *New York Times*; Ned Cramer, Braulio Agnese, Katie Gerfen, and Shelley Hutchins of *Architect* magazine; Noriko Tsukui with *A+U*; Verena Dauerer and Yariv Revah of *PingMag*; Sarah Rich of *Dwell*; Jori Erdmann and Jerry Portwood with the *Journal of Architectural Education*; Tyghe Trimble and Jen Barone with *Discover*; as well as Wendy Ju and Corina Yen with *Ambidextrous*.

I am also thankful to Tom Neilssen with the BrightSight Group; Susan Ades Stone with the World Science Festival; Johannes Salzbrunn with the Technical University of Vienna; Jose Luis De Vicente and Ana Fernández de la Mora with Arquinfad; Carol Derby with the Association for Contract Textiles; Mila Kennett-Reston and Mary Ellen Hynes with the Department of Homeland Security; Eric Letvin with URS Corporation; Mohammed Ettouney with Weidlinger Associates, Inc.; and Stephen Freitas with the Outdoor Advertising Association of America.

I would like to thank Kevin Lippert, Jennifer Thompson, Becca Casbon, and the other creative minds at Princeton Architectural Press for their ongoing support of this endeavor.

Most of all, I would like to thank my family for their loving support of this project.